NOTHING
SO
BROKEN

CHRIS RICHARDS

For information, contact

MSI Press, LLC
1760-F Airline Hwy #203
Hollister, CA 95023

Copyeditor: Betty Lou Leaver
Cover design & layout: Opeyemi Ikuborije

Permission granted for all individuals personally identified.

ISBN: 978-1-957354-61-3

Library of Congress Control Number: 2025906639

CONTENTS

For Larry and Steven

Acknowledgements

I'd like to thank Marie and Paul Brown, Janet DellaRocco, Bryant and Heather Richards, Denise Pachan, Margaret Pachan, and Regina Gibbons for their support, cheerleading and feedback since day one. Special thanks to Bryant for planting the seed on writing this story after a particularly difficult day. "So, when are you going to write a book?" Those words and the ensuing conversation took root.

I'm very grateful to Christine Bott for her patience, memories, and help with many of the important details of Steven's story for which I was not present. This would have been a half-story without her help.

I'd also like to thank the following family and friends for their patience and valuable contributions to the various sections; George Bott Sr., Fred and Pam Bott, Joe and Karen Bott, George Bott Jr., John Bott, Kristine Salsich, Terry Bernard, Tom Grabauskas, Greg Jacobson, Mike Remuck, Ron Dunham, Kevin Plante, Chris Nelson, Mike Kimbar, Peter Wayne, Tara Marano, Kevin Powell, Vito Simone, Cathy Stephenson, Allen Hoffman, Jack Dupre, Susan Lasewicz, Peggy Brown, June Soloperto, Dianne Brisson, Carole Chiras, Deb Ternove, Gustav Wind, Melinda Kelly, Ashley Green, Payton Rawls, Erin Clements, Miles Carmody, Alysha Metcalf, The Westboro Writers Workshop, Kaylee Lambert, Joshua Salsich, Noah Richards, Jill Giard, Sue Schroeder, Hank Miller, Karen McDonald, John Arthur Lee, Catherine Solcito, Jeff Lerman, and John Quinn.

I can never adequately thank Ethel Lee-Miller—coach, editor, and friend—for her insight, inspiration, and generosity. Her impact on this book and on me as a writer is immeasurable.

And last, a very special thank you to Steven Bott and Larry Richards for their courage and trust in me. Steven, thank you for allowing me to tell our stories together, teammates once again. Dad, thank you for allowing me to glimpse 1968 and the insanity that infected a nation. Thank you for keeping it at bay from your family.

Prologue

A week before the 1972 election, President Richard Nixon gave a wide-ranging speech to the nation, laying out his vision for the next four years should he be re-elected. Early in his talk he reviewed the record of his Vietnam policy and reported that "we have reached substantial agreement on most of the terms of a settlement" (Richard Nixon Foundation, 2017) with North Vietnam.

Nixon would go on to retain the Presidency in an historic landslide over South Dakota Senator George McGovern, carrying 49 of the 50 states and 96% of the electoral votes. Quite a contrast to four years earlier, when he squeaked by with only 56% of the electoral votes against Hubert Humphrey.

Shortly after Nixon was sworn into his second term, the governments of the United States, Republic of Vietnam (South Vietnam), Democratic Republic of Vietnam (North Vietnam) and the Provisional Revolutionary Government of the Republic of South Vietnam (South Vietnamese Communists) signed the *Paris Peace Accords*, or officially, the *Agreement on Ending the War and Restoring Peace in Viet Nam*. The agreement presented a multitude of terms, including the withdrawal of all U.S. and allied forces; the return of prisoners of war; a ban on introducing further military personnel into South Vietnam; a cease-fire in South Vietnam along with delineations of communist and government zones of control; and the reunification of Vietnam through peaceful means without coercion or annexation by either party, and without foreign interference. The agreement was negotiated by U.S. National Security Adviser Henry Kissinger and North Vietnamese Politburo member Lê Đức Thọ. Both men were awarded the 1973 Nobel Peace Prize for their efforts although Thọ refused to accept the award.

Two months later, President Nixon addressed the people of the United States:

> For the first time in 12 years, no American military forces are in Vietnam. All of our American POWs are on their way home. The 17 million people of South Vietnam have the right to choose their own government without outside interference…. We can be proud tonight of the fact that we have achieved our goal of obtaining an agreement which provides peace with honor in Vietnam (Richard Nixon Foundation, 2017).

Within weeks of Nixon's speech, open fighting broke out between South and North Vietnam and quickly escalated. The bloodshed would continue for two more years and conclude with the fall of Saigon.

Back in the States, we were reminded that wars never end with peace accords, treaties, and presidential announcements. Wars always end at the same place, with families.

I.

-how to wrestle with your father-

First, you will need a brother who enjoys conflict. He will make excellent fodder. Second, you will require a co-conspirator mother. As you crawl under the kitchen table, give her the *shhh* sign and wait for her nod that the coast is clear. Finally, you will need a low-lying lounger with a nonlinear profile in the living room (affectionately named the Black Bear) where your father likes to relax and read.

When your father is properly engrossed in his book, send in your brother like William Wallace at the raid of Scone. It is good if your brother is long and lanky, requiring both of your father's hands to contain. It is even better if he is armed with a pillow or stuffed animal. Once your brother is wrapped and wriggling and giggling, flank your father. If possible, climb onto the top of the Black Bear and leap. Notice how your father hardly moves when you land on his back, which feels as broad as a car hood. Focus on tickling his neck, because everywhere else will feel like tickling sheetrock. When your brother is worn out, your father will flip you off his back and tickle both of you, side by side. "Puppies have to play!" he will say, grinning. If he is unshaven, he may rub his scruff gently on your neck or belly, causing you to shriek. Your mother will watch and laugh and say she's not getting involved as you plead for rescue. If you have a dog, it will likely bark and wag its tail in excitement but do not expect any help. It may even happily lick your face.

When you finally escape, retreat to your bedroom to devise your next strategy. Notice how exhaustion complements joy. This will be important.

-in Vietnam-

My father said there were farms with pigs, noisy and smelly, and then he snorted, making my brother and me laugh. He could always make us laugh.

There was the time he was resting on the ground at night and something fast and heavy ran up to him and leapt off his chest, disappearing into the jungle. Too big to be a squirrel or cat, he said. He asked the other soldiers what it could've been, but nobody knew. They were strangers in a strange land.

Then there were the rats, or mice the size of rats, rummaging the barracks for food. If he ate crackers in bed, the next morning he'd awake to a rodent sitting on his stomach, sniffing for crumbs. So, my mother mailed him some traps. Fifteen dead rats in one night. Twelve the next. My brother and I squirmed. The soldiers gave the rats to the local villagers, who used them for stews. Papasan, the nearby farmer, was always grateful.

One morning the traps were empty.

No rats? What happened to the rats?

My father paused. My brother and I leaned forward.

"Cobras!" he exclaimed. Cobras had made a nest beneath the barracks. "Everybody out!"

The MP's gassed under the building and waited with pistols ready for the snakes to show themselves.

I checked under my bed for weeks!

-hero-

Long before I understood anything about Phu Bai and Lyndon Johnson and the Viet Cong there was York, Maine.

Our family made an annual summer trip to the popular beach town with Aunt Jan and Cousin Rob. Each July, an old, worn cottage, filled with mariner chachka and seagull figurines welcomed us for a week of sun and surf. We'd haul our towels, coolers, and inflatable rafts a half mile to Long Sands with its icy cold waves and striking dark grey beach. I forget where the sand came from. At night, the kids slept on the pull-out sofa in the living room in front of the small black-and-white TV. We'd drift off to sleep with the ocean scent spilling in through the windows.

Some evenings my father would drive us to the local amusement park for the rides, games, and video arcades. The Ferris Wheel would fill the darkening sky with a towering ring of lights. Children screamed as the roller coaster cars dove and roared along the creaky metal frame. Bob Seger and John Cougar Mellencamp blared through staticky speakers mounted on poles, and there were greasy pizza slices, Cokes, and an outside chance at fried dough if we behaved.

One cloudy afternoon, we decided to skip the beach and head straight for the amusement park. The place was empty. No lines. No waiting. We could try anything we wanted.

We sprinted for the Rolling Barrel, the preferred attraction for the older kids: middle schoolers and teenagers. The large, slowly spinning cylinder sat right next to the Funhouse. The teenagers always made it look easy, walking in place, as it rotated.

I ventured in first, cautiously. After a few awkward steps, I had the hang of it, my feet matching the speed of the barrel. Cousin Rob entered next. When the two of us were steady, my younger brother, Bryant, joined us. The parents stood at the entryway, enjoying our enjoyment, until somebody fell. I'm not sure who, but he tumbled into the two of us who were still upright, and suddenly the rolling barrel was a clothes dryer with three kids flopping around, smacking knees, shoulders, and elbows. Just as one of us would start to regain our footing, someone else would fall into him.

Ow!

Ouch!

Ow!

Help!

The parents' smiles vanished.

"Larry!" my mother yelled, and my father jumped into action.

He stepped into the barrel and lifted Bry up with one arm but then stumbled. Rob might have taken out his legs, I'm not sure, and down they went. For some reason, maybe for the ubiquitous parking meters throughout town, my father had what sounded like thirty dollars' worth of change in his pockets, and it emptied into the barrel. Nickels, dimes, and quarters rained down on us as we flopped and flipped. I'd started crawling to the other side when I felt a large foot give me a gentle push in the butt, dumping me onto steady ground. I turned to see my father lying on his back, Bry in his hands, Rob by his left hip, the barrel pulling all three and the mass of coins up one side.

"Larry! Larry! Larry!" My mother and Aunt Jan shouted.

My father somehow shimmied on his backside to the bottom of the barrel and out the front with a kid in each arm. Mom and Aunt Jan came around the side of the cylinder, flushed, wiping at tears, still laughing. We were a little sore, but Bry and Rob were already back at the edge of the barrel, trying to scoop up the money sliding around. My mother gazed up at my father lovingly, eyes wet. Our hero! Just like the movies!

-nemesis-

Most of my childhood mornings were marked by my father's absence. Running five miles every day, he said. Just something he needed to do, he said.

So, on weekdays before school, I'd wait at the kitchen window—teeth brushed, Space 1999 lunch box in hand—and watch the royal blue AMC wagon puff smoke from its exhaust in the Bott family's driveway two houses away.

My mother would run final checks with me—coat? gloves? hat?—from my brother's bedroom because "Defiant Bryant" loathed mornings. He started each day looking like he'd slept through a cyclone. Dressing him for kindergarten was like Greco-Roman wrestling with a grumpy squid.

When the old AMC wagon finally putted its way up the road, I'd book-it out of the house and wait at the end of our sidewalk. Mrs. Bott's smile through the window served as the weather barometer. An authentic grin occurred once a year, the first day of school, a tired or forced smile indicated trouble, and a frown would send my heartrate soaring.

Despite the sometimes-foreboding signs, I was always excited to see Johnny. Johnny Bott and I clicked from day one. Same age, same school, same classes, and an almost identical match in interests: Star Wars, Star Trek, Star Blazers, wiffleball, football, dodgeball, basketball, monster ball, tag, freeze tag, monster tag, blanket forts, tree forts, dogs, swimming, fishing, crayfishing, Atari, arcades, Creature Double Feature. We were Han and Luke battling stormtroopers. We were Godzilla and King Kong destroying Tokyo, or Fox Kid and Panther Kid protecting the neighborhood from Willow Walker, the evil river spirit. We ran all operations out of our secret base, high up in the Maple tree in my front yard.

Who gets to meet his best bud at age four? And live right next to him? Fate had gifted us something special!

But part of the deal was Steven, Johnny's older brother.

As I'd scooch into the backseat of the Bott-mobile, Steven would greet me with a cocky smirk, the diabolical wheels whirling away inside his head. Johnny would offer a brief smile and then stare out the window, counting the seconds.

"Hello, Christopher Robin." Mrs. Bott's standard salutation taught me early that if the Botts teased you, they liked you.

Pulling the car door closed, I knew the next few minutes were going to be the worst few minutes of my day. Johnny had four older brothers, but Steven was the closest to us in age, two years older, and Steven was... challenging.

One morning, as we merged into a busy intersection, Steven yanked my boot off with an evil giggle and tossed it into the way-back storage area of the car. Blood boiling, I clambered over the back seat to retrieve it while he went to work untying Johnny's sneakers. Johnny swore and cracked him with his Star Wars lunchbox, so Steven wrenched the box from Johnny's hands and dumped the contents onto the floor.

Mrs. Bott turned and swiped at him with a rolled-up magazine, "Knock it off!" she yelled as Steven dodged and ducked, cackling like a deranged elf. Once past the intersection, she pulled over to the side of the road.

"Get. In. Front. Now!" She said *now* without opening her mouth, a tremor in her voice, the tendons in her neck pronounced. I wanted to jump out of the car.

Steven hopped over the seat and landed on top of a small pile of newspapers and magazines, knocking them onto the floor. I pressed up against the back door while Mrs. Bott formed complete sentences with her mouth shut, teeth gritted.

With five boys between the ages of eight and sixteen, Christine Bott lived every moment pre-occupied with the certainty that something bad (and likely expensive) was happening somewhere. There was an endless list of damage reports to review, assess—sigh!—and respond to.

As for Steven, the rides to school were merely the opening acts. I can remember one rainy morning during indoor recess the gymnasium smelling like wet dog. All of the grade three, four and five students sat in the bleachers under the watchful eye of Mr. Rogers, the Vice Principal and Mr. Kerins, the gym teacher. All, except two. Behind the adults, on the opposite side of the gym Stephen Ceccarini chased Steven in and out of a series of doors like in the Scooby Doo cartoons. Moments before, Steven had given Ceccarini the finger while neither teacher was looking.

Ceccarini never caught Steven, but Mr. Kerins caught Ceccarini and sent him to the principal's office. I don't remember how Steven avoided a similar fate, but I do remember him beaming with how things played out— until Charlie Darling, a tall, athletic kid ambushed him from behind one of the doors. Charlie dragged Steven back into the hallway and pounded the daylights out of him. One of my happier childhood memories.

Thankfully, I didn't see much of Steven during school hours, but per Mrs. Bott, he often got in trouble for doing "stupid stuff," like rubbing up against the blackboard to erase class assignments or using reflective surfaces to shine points of sunlight onto the back of the teacher's head. Whenever Mrs. Bott saw her son sitting outside next to the mailbox, she knew a disciplinary letter was on the way.

After school was more of the same.

"Knock it off, or I'm stoppin' this bus!" the driver would shout at the fifth graders in the back seats where Steven sat.

"Stop it right now, or I'm calling Officer Perry!" Mrs. Butler, the poor harried crossing guard yelled at Steven (and others) once the bus dropped us off and the fights began.

Knock it off!

Cut it out!

Cut the shit!

Leave me alone!

I'm telling Ma!

He never stopped, possessed by some demon that needed to provoke others. When his crosshairs found me, which was often, I'd end up rattled, bruised, and teary-eyed. Oh yes, and furious. His constant attacks stoked a deep visceral rage within my young soul. Nobody bullied me like him. I'd pray at night for strength to fight back. I'd pray that he would no longer be Johnny's brother. I'd pray that he would go away and disappear.

But God didn't seem to be listening.

-mayhem-

One of the most frustrating things about Steven was also one of the most impressive things about Steven. He was a gifted athlete.

I remember an autumn afternoon, he had me pinned to the ground, sitting atop my chest, knees digging into my arms as he pounded away. The blows struck my eyes and cheeks like padded hammers while I squirmed and launched my hips upward, trying to buck him off. When enough was enough, he jumped off me and back into the mayhem of some made-up game involving a football. I stumbled toward the house, crying, certain that Steven had knocked one my eyeballs out of its socket.

Kids were chasing other kids around the yard, and my eye must have been lying on the grass somewhere. I hoped nobody stepped on it because then I'd be blinded for life. I'd have to wear a patch like Snake Plissken in *Escape from New York*. My father wouldn't let me watch the movie with him, said it was too violent, but I snuck out of bed that night and peeked in from the darkened hallway and saw the President's severed finger wrapped in cloth. Not sure how I slept after that, but I knew Snake would save the day. Maybe, it wouldn't be so bad to wear an eye patch.

Plus, the beating was totally worth it. It was retribution for an airborne lariat where I tried to decapitate Steven. I'd seen the move dozens of times on Saturday morning wrestling with Rowdy Roddy Piper and the Iron

Sheik, but Steven never saw it coming that afternoon. He'd been focused on catching Johnny, who was scrambling with the football in hand past the overgrown forsythia bushes toward the 80-foot, high-tension wire tower. I leapt toward Steven as he gained on his brother, my arm straight out to the side. The bend in my elbow cradled his throat, and our momentums did the rest. Glorious.

Clean shots like that never happened to my nemesis. He'd stiff-arm, twist, and duck his way out of any solid hit. Even the older, stronger kids—Jeff and Neil Roberts, George Bott, Danny Gilbert—had their hands full with him in fights and athletic competitions. In a neighborhood overflowing with active boys, something about him shone differently.

I'll never forget the football games. I'd be latched onto his ankle like a bear trap, trying to slow him down and hoping a teammate would pop him as he dragged me toward the end-zone. But it never happened. Bry, Johnny, Cousin Rob, and Jeff Gilbert bounced off him like I did. Big Ray Gauthier was strong enough, but Big Ray was slow. Steven wasn't, even with me attached. That was the problem; he was just as tall as the other big kids, but he was lean and fast. Big kids were supposed to lumber.

Other sports invoked similar frustration. He turned on fastballs and sat on curves in wiffle ball, sending home runs sailing over our neighbor's split-rail fence. He hustled singles into doubles, doubles into triples, dodging and diving as we tried to hit him with the ball to get him out. He'd often taunt us into throwing it at him and then take off for extra bases when we missed. "For God's sake, don't throw it!" we'd yell at each other.

Sniffling, I shuffled up to my mother at the kitchen table. She was engrossed in conversation with Mrs. Bott, and they both seemed happy and relaxed. A brief respite from their boys.

They turned to me, their smiles disappearing, and my mother conducted her inspection; teeth intact, nose straight, no blood.

But Mom, my eye!

She frowned and pulled me closer.

"It's just a little red, honey," she said. "You'll be fine." And she sent me outside, back into the ring.

-home run-

Coach Shepard, or 'Shep' to most, wanted me to do five push-ups every night during the commercial breaks of my favorite TV show, which was probably *Three's Company* at that time. It was my final year of little league and I hated it. I wasn't on Johnny's team with Mr. Bott coaching. They weren't any good, but they had fun. No, I played for the Millbury Credit Union juggernaut with the hyper-competitive coaches. We won all the time, but I didn't smile much. And my parents had to talk me out of quitting on several occasions.

I also struggled with the weight of the bat, tapping slow and sad ground balls to first or second base for easy outs. Then, Shep assigned the push-up homework.

Toward the end of the season at East Millbury Park, another victory well in hand, I stepped up to bat against Paul Norman, a good friend. Early in the count, the ball got away from him and zipped in high and tight, sending me diving into the dirt. After brushing myself off, I dug back into the box, a little fire in my belly, fingers tightening around the bat. Five push-ups.

I hit his next pitch with the "sweet spot" of the barrel, barely feeling the impact, and launched the ball into a high, lazy pop fly that cleared the fence in left-center field by a few inches.

I can still hear Shep's gravelly voice among the cheering, the infield dirt crunching under my cleats as I held onto my helmet to keep it from wobbling off my head. Rounding second base, I faced my teammates and coaches on the bench and the parents in the bleachers behind them.

A standing ovation!

For me!

Off to the right, up on the hill in a lawn chair, my father sat alone with his nose buried in a novel. Probably Heinlen or Robert E. Howard. He went to most games but never sat with the other parents, always off to the side and usually a few feet behind the crowd. He enjoyed his solitude. He enjoyed his books.

As I approached third, he looked up and processed my slow jog around the bases.

"Heyyy!!" he yelled and shot his hands straight up in the air. "Home run!!"

My body felt weightless.

I'm sure I jabbered to him in the car all the way to the Ice Cream Barn after the game, the vanilla soft-serve cone and chocolate jimmies shutting me up for maybe ten minutes.

-basketball-

Steven and Johnny sped behind the back of the Bott house, past the above-ground pool, then circled round the front yard. Like a gazelle evading a mildly concussed cheetah, Johnny zigged and zagged, keeping his brother off balance just enough to maintain a small lead. Steven's face blazed, a contorted mask of rage, a small red circle on his forehead from where the ball had struck moments before.

The boys took a sharp turn into the garage and disappeared, the screen door slamming behind them. If my window was open and the nearby highway wasn't busy with rush hour traffic, I could hear that screen door slam all the way from my bedroom. I would often look out the window to see who was exiting the garage.

We named the Bott's driveway, the Bott-on Garden, even though it resembled a miniature golf course more than the Boston Celtics' home court. It sloped 60 feet up from the road, taking a sharp right to the single car garage on the first floor of the house. The basketball hoop stood on the edge of the turn, facing the garage. Behind the basket, scrub tapered into a wooded undeveloped lot, while off to the right of the basket a steep hill covered with thorny undergrowth descended into a brook.

We didn't lose the ball in the water *every* time we played, but one errant bounce off someone's foot shot it down the hill with a pack of screaming kids in pursuit. Johnny or Steven would hop across our makeshift stone crossing at the stream's narrowest section and follow it along the opposite bank while the rest of us threw rocks at the ball, trying to knock it over to them. Time was a ticking; downstream the brook fed into the wider, deeper, and swifter Blackstone River. Somewhere in Rhode Island a kid ended up with about 20 free basketballs.

Of all the sports we played as children, basketball became everyone's favorite. We bounced between soccer, baseball, and basketball in the seasonal town leagues, but cheering for the Red Sox back then bordered on masochism while baseball involved a lot of standing around. Soccer required exceptional foot skills, which none of us possessed. Basketball though—basketball was fast-paced for our fast minds and energetic bodies, and the Celtics with Bird, McHale, and Parish contended for championships every year.

Steven, of course, excelled on the court. He was built for the game: tall, long and quick. He was also clever and dexterous, dribbling with either hand, behind his back, between his legs. He could pass to one teammate while looking at someone else. But alongside his jaw-dropping skills came his never-ending taunting and arguing.

That afternoon we'd started with five players, but George, Johnny and Steven's older brother, had become so fed up with Steven that he quit halfway through the game.

I liked George. He walked and talked like a Bott but differed in many ways from his siblings. The middle brother and lone introvert of the family had interests ranging well beyond the playing fields and gymnasiums and possessed a strong connection with animals. He was also five years older than I and in high school, so it was wicked cool when he played with us.

Instead of us rebalancing into teams of two players with George gone, Steven tried to beat Johnny, Bry, and me—one against three. Pure insanity.

He might have pulled it off, though, if not for a single play; Johnny rushed in on Steven's blind side just as Steven brought the ball up to shoot a layup, knocking the ball out of his hands and off his rising knee. The ball ricocheted hard off the basketball pole and back into Steven's groin. He let out an *oomph,* grabbed his privates, and fell over sideways. Another one of my childhood highlights.

After the giggles faded and the air returned to Steven's lungs, the brothers got into each other's faces—*you fouled me, respect the call, my ball, get your hands off me, I called it, give me the ball, stop pushing me*—and before I knew it Johnny was rearing back with the ball like Nolan Ryan.

That left Bry and me alone in the driveway, wondering if we should stay or head back home. Three years apart in age and worlds apart in appearance, my brother's blue eyes and blonde hair juxtaposed with my dark features. Adults gave pause when we stood side-by-side, thinking we were pulling their leg. Bry was also growing at an accelerated rate, allowing him to play physical games like basketball with older kids. He passed me in height when he reached seventh grade and passed most of the town's population when he reached ninth.

Before we could decide what to do, a shrill scream erupted within the Bott house, and the screen door flew open. Steven, red-faced, hair a wild mess, shot out of the garage with George right on his heels. They beelined past us into the woods without a word.

Moments later, Mrs. Bott hurried out into the driveway, hands on hips. I may be embellishing, but I think she was muttering something about how two children would've been enough.

-fracture-

When we first moved into our house, there were four maple trees in the front yard, two bigger and two smaller, and I thought, *how nice, just like us.*

Then, there's my mother turning off the television and sitting down on the couch to my left. I don't recall the day of the week, or the weather, or what show I was watching. But I remember she was to my left.

My parents had begun their journey together in 1962, 16-year-olds slow dancing at a house party. My mother said my father was funny and outspoken although he could be embarrassing at times. They graduated from Sutton High School among a class of 50 students in 1964, a few months before North Vietnam attacked two U.S. destroyers in the Gulf of Tonkin. Three years later they were married.

They presented few symptoms of discord outside of sleeping in separate rooms. Dad was a restless sleeper and didn't want to disturb Mom, we were told.

"We had 11 good years," my mother would say later. "And five tough ones."

Two weeks after the talk with my mother, we were standing on the porch, watching my father drive away in a dark brown passenger van. I remember the *how-the-hell-did-we-get-here?* look on his face as he waved goodbye. My mother stood between Bry and me, her arms around us as we waved back. We kept waving long after the van disappeared.

-man of the house-

So, life became table for three diners and an empty chair, and my mother's full-time job, and my father's apartment in the city, and a half-mowed lawn, and a flooded basement, and pitying looks from Mrs. Bott, and a 9-year-old brother, and sixth grade social studies, and the guidance counselor, and dark spots on the walls, and dogs that dug up the yard, and ants crawling inside my stomach, and weekends away from friends, and wet leaves that killed the grass and Parents Without Partners Dances, and *Nobody ever asked me what I thought,* and therapists, and a driveway buried in snow, and big angry kids on the bus, and fights and scrapes and bruises, and empty flower planters, and a rusting shed in the backyard, and running home after school, and a hidden key on the porch, and a Louisville slugger in the closet, and a bottle of Jack in the woods, and a suspected drug dealer in the single-wide pink trailer across the street, and *make sure the door is locked until I get home!*

-ally-

The largest maple tree along with the secret superhero base was cut down first. It was deemed too big and much too close to the house; the roots had expanded, building up pressure against the foundation.

A year or two later, the pair of smaller maples followed, but the landscaper left the stumps and never came back to finish. They sat anchored in the front lawn for years, oozing rainbow-bright rot into the surrounding grass. Mushrooms and other fungi blossomed.

But Kim Anderson was wicked mint, an eighth grader with chestnut curls framing freckled cheeks and soft brown eyes. Her friends, Marie Labreck and Michelle Nadeau, were also suddenly cute, and then there was this tall girl in my math class with blond, spikey hair, and *Oh My God! What was happening to me?*

I'd notice Kim talking with her friends in the cafeteria, and my cheeks would flush. She'd walk by me in the hall, and my pulse would pound like I was running sprints. I think she might have smiled at me once, and the afternoon went blank.

Fortunately, I had Johnny, resident Love Guru, to consult on such matters. With four older brothers, he possessed an impressive understanding of girls and dating. I needed to ask Kim out, he said. Out where? Outside? "Out on a date," he said, shaking his head. We agreed that outsourcing would be wise. Johnny would ask her for me.

The day of the question arrived. Was Kim to be my girlfriend? Were we *going out*? God help me if she said yes. I left school early that day for a doctor's appointment and missed Johnny on the bus ride home. Later in the afternoon, I walked down to the Bott's house.

Steven answered the door. Everything inside me tightened. "Is Johnny home?" I asked.

He turned and yelled upstairs for John while I rocked back and forth, wishing he would go away. Instead, he waited, a curious expression on his face.

"I heard you asked Kim Anderson out," he said.

I stopped breathing.

"Nice move," he said.

A compliment from Steven; the Apocalypse would be starting soon.

"She's cute," he added.

"Yes, she is," I said, and suddenly, if only for a moment, I had an older brother.

Granted, he was Johnny's brother, but Johnny had three other brothers he preferred to Steven. For me, though, the oldest of my generation, there were no brothers, sisters, or cousins to consult. Steven, a freshman in high school, had been on dates and knew the difference between a *nice move* and a bad one. His validation calmed the confusion I felt. Maybe everything would be all right with this dating thing.

Johnny appeared at the doorway with a couple of gloves and a baseball. "So?" I asked.

He handed me one of the gloves and joined me outside. "Well, she has no idea who you are ... but she's definitely not interested."

"Oh."

Steven paused before shutting the door and shrugged. "It was still a nice move."

-flu-

About a year after the divorce, my father came down with flu-like symptoms during the summer. He missed three days of work before dragging himself back into the office. He was freezing in July. He was freezing in August. He stopped his daily runs and took long walks instead, his energy half of what it used to be.

I have no recollection of this. I imagine my brother and I skipped our weekend visits during this time. I imagine my parents made up a story about why he couldn't see us.

-mentor-

Backyard basketball games end when a team scores 11 points. Or seven. Or 15. Or 21 if no one else is waiting to play. I don't know who came up with the rules or why the fascination with odd numbers, but these finish lines have been used for as long as I can remember.

For the entire summer of '85, I played one-on-ones to 100. It began with another Steven-versus-everyone moment. A group of kids, maybe ten of us, converged in the road in front of the Bott house, some on bikes, some on foot, all of us deciding what to do with our afternoon.

Our neighborhood consisted of 20 homes along a half-mile cul-de-sac nestled alongside interstate Route 146. There were above-ground swimming pools, uneven yards of all sizes and shapes, two wooded areas with trails, and the winding, little brook behind the Bott's. And so many kids—boys mostly, growing like wildfires. Energy everywhere. Dirt had to be dug. Races run. Bikes crashed. Trees climbed. The woods and river had to be explored for snakes, frogs, salamanders and snapping turtles. And rocks had to be thrown into the river. The larger the rock, the bigger the splash and better the chances of dousing a fellow adventurer.

That fateful afternoon, Bry and Johnny were also there deciding what do, with most of us interested in exploring the woods and biking along the trails.

Not Steven. The athlete wanted to play a game—wiffle ball, basketball, football, anything. Figuring some of us would be on the same page, he started off toward his house.

Nobody was on the same page. When he realized he was alone, he turned to me, of all the kids, asking what I was going to do? I sat on my bike; the trails were calling. What did he think I was going to do? But something in his voice, a wisp of loneliness or maybe just him standing by himself made me hesitate. Then, I made the choice that changed everything. I mumbled something about playing ball, not quite believing my own words.

Johnny's eyes went wide and gave me a look like I was insane. The gang bolted, leaving me alone with Steven. Two kids and an empty afternoon to fill. The one-on-one marathon was born.

The contest took forever, but each basket had little to do with the final score, so we didn't argue. It was surprisingly pleasant, like a normal day with a normal kid. We played again the next day. And then the following week. And then the week after that and so on. The days warmed and stretched. We lost a couple balls to the river, and I lost every single game. But once again, I had a big brother, challenging me and there's no better teacher than a challenge.

A single goal drove us through the marathons: high school tryouts. Come winter, I'd be a freshman, with the junior varsity team in my sights and Steven would be trying out for varsity as a junior. The previous year he'd started for the JV team, while Johnny and I had won the Junior High town league, coached by Joe Bott, second oldest of the five brothers.

One particularly oppressive July afternoon, both of us exhausted, Steven suggested we get a drink. We hiked upstairs to his kitchen, a furnace with windows, and he pulled out a pitcher of iced tea from the refrigerator. For reasons I've long forgotten, the Botts never used air conditioning, relying on the pool during the day and fans at night. Mr. Bott and the boys also went shirtless from June to September and Mrs. Bott didn't seem to have enough time to bother with the heat. A box fan always sat on the countertop, circulating the hot air like a hair dryer.

I believe that was the first summer Steven and I ever talked like friends talk—probably about how the Celtics were robbed (they had lost in the Finals), the Lakers were bums, and oh, how we hated Kareem, Dire Straits, *Money for Nothing, Back to the Future,* and *Rambo* ("Murdoch … I'm coming for you!"). He probably teased me about Kim and probed about other crushes, but I never dated anyone in junior high, still too tentative.

At some point, Joe Bott strolled into the kitchen, wearing a bathing suit and flip flops, a towel draped over his shoulder. Of the five brothers, Joe was the most affable, possessing an effortless and infectious smile. At 21, he was already a natural with kids and an enjoyable coach.

Who's winning, he might've asked. Steven's cheating, I likely answered. He told me he'd run into Terry, who had asked if I'd been playing any ball this summer.

I choked on my drink. Terry was the JV coach. Asking about me.

Joe said I had a good chance of making the team. Just keep practicing. *Awe-some!*

Joe wasn't sure about Steven's chances though. Varsity was stacked, and Coach Dunham might not even want Steven, he might have to stay on JV. Joe winked at me as he said this. Juniors like Steven didn't play on JV. They either made varsity, or they were cut.

"Let's go. Right now!" Steven said. "I'll show you who's JV!"

"Sorry, I'm busy," Joe said and headed out the back door toward the pool.

I spent more time with Steven that summer than the previous ten combined. A mutual respect was forming, and his antagonistic behavior had softened from bullying and taunting to teasing. Johnny was still my best friend, but he was heading off to a vocational high school in the fall. He would make new friends. Take different classes. Ride a different bus. Time for me at the Bott home became split between the brothers, sowing confusion among the rest of the family.

"Oh, hello, Christopher," Mrs. Bott would answer the door. "Here to see Johnny … or Steven? Never mind, they're both home. Go on in and find whoever."

When Johnny and I became friends as 5-year-olds, a deep, simple bond formed between us, pulling on our families like magnets—rides to school and practices, sleepovers, cookouts, birthdays—and our parents grew close. When Steven and I became friends, the connection strengthened, and boundaries blurred. Mrs. Bott started referring to me as *son number six*. George would visit us for hours, talking to my mother and playing with our dog. Or our dog would visit George, disappearing into their house for an afternoon. Sometimes, I'd play at the Bott-on Garden by myself, chatting with Mr. Bott while he worked under a car. "Whatcha workin' on?" never really understanding his answers. He was always fixing something. Sometimes, I'd see my mother peek her head out of the porch door, looking around the yard for me, and I'd yell and wave to her. She'd wave and go back inside.

-wolverine-

My father moved again, this time into the duplex beside Aunt Jan and Cousin Rob in the nearby town of Sutton. A wonderful move for everyone. Rob, an only child, had family to play with on weekends, and Bry and I had someone to play with in an unfamiliar neighborhood. The adults could take turns watching us. Win-win-win.

My father gave Bry and me the larger bedroom upstairs, twin beds with our own bookcases and closets. The living room had a couch, a recliner, and a giant blue bean bag chair for us to wrestle on. The orange rugs didn't match the couches or the recliner, which had a lot of muted earth tones, with the occasional misplaced pop of color, but everything worked somehow. We loved it.

There was also a 4-foot-wide dark maple bookcase with all of my father's wonderful science fiction and fantasy novels: MacDonald, Heinlen, Hubbard, Herbert, Tolkien, Howard, Bradbury. My father told us about *Glory Road* and the indestructible Iggly. He talked about *Time Enough for Love* and how a man's brain got placed in a woman's body. He talked about Michael Valentine Smith and Grokking. I'd study the artwork on the covers—Conan swinging a two-handed battle axe at a gigantic snake— trying to imagine the story. By that time, I'd already read *The Lord of the Rings* and *The Chronicles of Narnia* and a good part of my junior high years had been spent in basements playing Dungeons and Dragons with Johnny, Bry and Rob.

I don't remember when I first noticed the action figures. I think it might have been Wolverine, black and yellow with pointed ears and adamantium claws on top of the refrigerator. Then there was Han Solo, perched on a windowsill in the kitchen, and three plastic green army men on the bookshelf in the living room. I'm not sure if I asked my father or if he just told me the story about his childhood, sleeping in a room that had a door to the creepy, unfinished basement. His mother had told him the monsters in the basement were his imagination, but if he believed in the monsters then he could have friends to protect him and fight those monsters. Together, he and his mother set up toy soldiers that faced toward the basement door, ready for action. He slept soundly.

Sutton is a small, mostly rural town amid the hills of central Massachusetts, maybe 15 minutes south of the city of Worcester. There are plenty of farms and maybe one or two stop lights. A policeman once joked to me that he was trying to solve his first murder case, a deer had been hit by a pickup. When I followed Wolverine's line of sight out the back door, it led into the spacious yard shaded by the gigantic oak under which we played football and wiffle ball. To the right, our neighbor's above-ground pool and splashing kids. To the left, an empty field of tall grass for hide and seek.

Han Solo peered out the window at the duplex next door and the army men watched the unpaved driveway and old ranch house where the Gravels, a sweet, elderly couple resided.

My father was a smidge under six feet with the shoulders and chest of a gorilla. He'd also been trained for military combat. Yet, every entryway into the apartment was being guarded.

From what? I wondered.

-eye of the storm-

The summer I turned 15, I started taking walks. They began on Sundays; I didn't want to be at my father's anymore. Friday nights were fun— Big Macs, vanilla shakes, and Miami Vice. Then, Saturday afternoons at the Playoff Arcade or basketball at the Main St. YMCA in Worcester are some of my happiest memories. But Sundays were different: trapped in an apartment that shared a muddy driveway with an old ranch and dilapidated garage. There was a liquor store up the road and two plastics factories across the street. It was all stupid.

I didn't really want to be at my mother's, either. The neighborhood had changed. The woods, once a haven for adventure, held secret supplies of cigarettes and beer. Steven could drive and had a summer job, so he disappeared. Johnny had new friends from his new high school and at least one girlfriend.

On Sunday mornings, all this boiled over, revealing life to be unfair and pointless and forcing me to do what most teenage boys would do— walk three miles to church.

I'd wake early, the apartment quiet with my father already at the Grafton Flea Market, hunting for the best deals around, and Bry out cold, still a hard sleeper with all that growing. At the end of the muddy driveway, I'd take a right onto Depot St. and follow it downhill to a small bridge that crossed the same basketball-pilfering Blackstone River that flowed behind the Bott's house miles away. After the bridge, a set of railroad tracks ran alongside the river and snuck behind the Polyvinyl Films plant. Leaving the road, I'd follow the tracks toward my hometown, Millbury.

A half mile into the journey, railroad met river as the Blackstone turned sharply north before curving west again. A seemingly forgotten and uncared for bridge straddled the water, waiting for trains I hoped no longer traveled that way. Massive stone slab bases held the beams, ties, and tracks in place, but rotted wood and crumbling stone belied the structure's apparent strength.

With my head down and heart pumping, I'd cross slowly. At the center of the bridge, a man-sized gap in the ties revealed a dizzying view of the water and rocks 30 feet below. A flash of vertigo and then dark thoughts ... *What if I fall? Nobody knows where I am. Nobody will figure it out in time.* Sometimes, a strong gust would make its way along the river, rippling the water, stirring the leaves, as it rushed toward me. Bracing myself, I'd lean into the wind and inch forward.

That first step onto the other side was the highlight of each week. The crunch of stone beneath my feet felt like a small miracle, my heart swelling with Hallelujahs.

A hundred yards later, the railroad shortcut ended, and I jumped back onto the road, all the way to the St. Brigid Church in the center of Millbury. For the mostly empty 9:00 Mass I sat in the back of an older crowd, unsure if I belonged, but feeling safe amid the old routines. My father had left Christianity years before, unsatisfied with its answers to his questions. My mother was still a regular but often went to a different Mass.

Either Father Kelly, a portly man with a gentle presence, or Father Markey of the silver hair and icy stare led the services. Father Markey's frustrations often spilled into his sermons: Where is everybody? Mass is every Sunday, not just Christmas and Easter! I always wanted to yell out, "Why are you complaining to us? We're the ones who are here!"

Instead of returning to my father's apartment afterward, I'd walk to my mother's, which was close to the church. If the weather was nice, my father would load up Bry and a carload of cousins and friends to bring to his parents' home on Lake Singletary. There was a Sunfish sailboat, rowboat, motorboat for waterskiing, fishing poles, a dock for cannonballing, picnic table for card games, a hammock, miles of woods in every direction, and two sweet, loving grandparents who lit up every time we walked in the door. All of this was ten minutes away; we'd been going there every summer since I was born.

I was no longer interested. My father would call or stop by my mother's house to invite me, but he never forced me to join them.

"Let me know if you change your mind," he'd say softly.

"I won't," I'd mutter.

And I didn't. After lunch, everybody headed for the lake. I headed for the highway.

The Washington St. Park lay on the other side of Route 146, a quick jog from my mother's house if you could dodge the cars. The park consisted of a basketball court, a small playground with swings and a spiral slide, and arguably the nicest baseball field in town. On summer nights, after the rush hour traffic, the ping of aluminum bats and cheering would carry throughout the valley into our backyard.

The basketball court lay at the farthest corner from the highway, nestled like an afterthought along the edge of the woods. I'd find the park empty, peel off my shirt and dribble slowly up and down along the sidelines; fingers searching for the gentle grooves in the ball, skin soaking in the sun, muscles lengthening and warming with blood.

Then, I was off. I'd loft the ball in a high arc toward the other end of the court and sprint after it, snatching it in front of the basket and flipping it upwards. Then I'd race in the other direction, pounding the ball into the pavement. Another wild shot. Sometimes, I'd collide with the support poles and pin-ball off them, circling back. Instead of picking the ball up, I'd kick it gently toward mid-court, the rules of the game slipping away. Ghostly defenders lunged at me as I weaved between them. Another full-court sprint. Back and forth and back and forth, and then I'd stop under the basket and bounce on my toes, reaching for the net. Twenty, 30, maybe 50 times. Eyes stinging with sweat. Lungs on fire. I'd pick up the ball and walk to the free throw line, steadying myself. Three dribbles. An exhale. *See it.* The net would pop up through the rim as the ball dropped in dead center.

Then more running and chasing and jumping, and the clattering, noisy mind fell into the background. Swirling emotion transformed into physical energy, burning itself out, and somewhere deep within that hurricane lay the eye. Something in the center harbored a clarity, a calm. I'd found the *me* I remembered before acne-speckled skin and a cracking voice, before two homes, girls, bullies, drug dealers. I could access it every time I played out the furious dance. Within movement lay stillness.

Hours later, I'd lie under a tree, sweat-soaked, the cool grass tickling my skin, the discomfort and moodiness receding. The sun peeked through the leaves, winking at me. *Welcome back*, the breeze whispered. The peace stayed for a day, maybe two.

The routine started on Sundays but soon spread into Saturday and the weekdays. Weeks and months were consumed. The summer disappeared. All that running and nowhere to go, until one afternoon when Steven pulled up at the end of my driveway in Joe's white Mustang.

-summer league-

"Do you have a maroon shirt?" Steven asked.

I did have a maroon shirt.

"Good," he shifted into park. "Grab it, and get in."

Despite Joe's teasing, Steven had made the high school varsity team as a junior the previous season. And what a team! Three senior all-stars, a versatile junior at forward, and a tall, bruising sophomore at center. There were also a few players on the bench, like Steven, who could contribute when needed. Speedy, athletic, and skilled—a buzz grew around town.

In a twist of fate (and feet), the starting point guard broke his ankle during the second game of the season. Eight-week recovery. Steven was the next man up, and it was a hell of a jump. Memorize the plays, understand teammates' tendencies, and recognize opponent defenses STAT! All while playing in a game bigger and faster than anything he'd ever experienced.

My crazy, trash-talking friend quieted down (temporarily), and within a few weeks he was running the offense, limiting his mistakes, and scoring the occasional basket. With Steven at the point, everyone else focused on tearing up opponents' defenses. The result was an historic success: the State Championship game, losing by three points to a heavily favored Boston school in a tense contest after which the Boston coach chased one of the referees off the court.

"We don't have enough guys for tonight," Steven said as I settled into the passenger seat. Everyone was away on vacation. I was suddenly on the team. What team?

"High school varsity," he said.

Varsity?

"We're playing in the town men's league this summer."

Men's league?

Uh, oh!

As hoped, I'd made the JV squad my freshmen year, starting the season on the bench. Minutes were earned, hustling and listening to Terry. I was coachable, if nothing else. I was also maybe 5'7", 125 pounds, not exactly built for men's basketball. But for one night, varsity!

Steven broke several speed limits on the way to the park, and we arrived just as the referees were asking the other players about a forfeit. Our teammates eyed me suspiciously.

"He's with us tonight," Steven said. "Don't worry, I taught him everything he knows."

They laughed and thanked me for showing up. We had five players. No substitutes. I would have to play the entire game.

Our opponents, a team of fathers and uncles in their late twenties, practiced at the opposite basket with their beefy arms and broad shoulders. We would be countering with wispy mustaches, feathered mullets, and pimples. Ron Dunham, the high school varsity head coach, believed that having the varsity team play in a summer men's league would toughen them up for the upcoming season.

A few minutes into the game, I found myself wide open on the wing. Steven threw me a hard pass while pointing at another teammate. If I hadn't seen that showboating, no-look move of his at the Bott-on Garden, the ball would have hit me in the head. Instead, in one fluid motion, I caught the pass, let it fly, and we were up two points.

From that moment on, I could do no wrong: the point guard for the other team dribbled the ball off his foot right into my hands. An uncontested layup. A long pass from Steven for a fast break basket. A jump shot on the baseline. A bullet pass to 'Grubby' (Tom Grabauskas). Every play the ball found me, and every play something good happened.

By halftime, it was a rout, and we were all scratching our heads and grinning. I hadn't just stepped in as the fifth man, the JV scrub to prevent the forfeit. I was leading us. Scoring. Steals. Loose balls. Assists. Never had I dominated a game like that in my life. And then to do it in the men's league. How?

On the car ride home, Steven asked me to play in the next game, too. Of course, I would, but could I play that well again?

The rest of the games that summer all resulted in solid, if not impressive, performances. I played as if I belonged, as if I had been on the varsity since day one. By the third game, my presence had been accepted and expected by upperclassmen I hadn't known two weeks earlier.

Terry attended one of the games, approaching me afterward. Seven months had passed since our JV season. By the end of the year, I'd scrapped my way into his starting lineup but was far from being one of his top players.

"Nice game," he said. "Been playing a lot?"

I went on about camps and clinics and parks and practicing and how Steven showed up one day and asked me to play in the league and wow and on and on and on.

He chuckled. "Keep it up, and you may be doing a lot more of that come fall."

I responded with a smile that almost split my head in half. He was going to tell Coach Dunham, and Coach Dunham was going to tell him that he was out of his ever-loving mind. But I might actually get a tryout for a team that had just made it to the state championship.

-dreamwork-

Stomach cancer.

The pain ripped through his abdomen, a hungry fire, melting his insides and radiating outwards. He clenched his belly, holding it tight as it swelled, threatening to burst. The heat worked its way into his chest and legs, devouring everything, slowly turning his cells to rot...

When my father woke from the dream, his sheets were soaked with sweat. He showered, went to work, came home, and washed his sheets. The next night he dreamt the same dream.

And then again.

"Three whole nights?" I asked. A 5-minute nightmare could feel like an eternity.

"Well, two and half," my father said. "I died during the dream on the third night."

Each morning, he journaled the dream in a notebook by his bed. I remember a notebook (but can't picture the color) on the floor by the nightstand. He told me that he often journaled his dreams, looking for messages or teachings. I don't think I ever read it or even peeked. I'm not sure why a curious teen wouldn't want to know, but something about him always required a certain distance. It wasn't fear; it was mostly respect. Mostly.

-practice-

I was going to throw up.

Steven drove us to practice while my insides tumbled and flipped. I focused on the dashboard, wondering if I was about to cover it with lunch. Mrs. Bott would kill me.

Steven assured me I'd be fine. These were the same guys from the summer league.

Yes, but Coach Dunham...

Ron Dunham was a no-nonsense tactician who cranked out winning teams like a factory. My previous two coaches—Joe Bott, a good-natured jokester, and Terry, a wellspring of encouragement—focused on teaching first, winning second. I'd seen Coach Dunham briefly at a camp over the summer and felt like I was under a microscope.

"Don't worry about Dunham," Steven added. "He doesn't know what he's doing."

Coach Dunham would later be voted into the Coaches Hall of Fame, but Steven's misinformed pep talk still helped. The thought of Coach and Steven together, the machine and the meatball. How did they coexist? How did Coach not kill him?

Practice started with a short lecture as I stood on the outskirts of a huddle engulfed in a cloud of body odor strong enough to asphyxiate a ferret. All the players had hairy armpits and muscles and at least three inches on me. Grubby's five o'clock shadow shone dark and scruffy, even though it was only 2:00. Who the hell were these gorillas? Where was my summer league team? I didn't belong. I should have stayed with my classmates on JV.

At some point, Steven caught my eye, winked, and let loose a fart so loud, it was a miracle his shorts didn't fly off.

Coach glowered as the group stumbled into nervous laughter.

"Sorry, Coach, must be the cafeteria food," Steven said.

Coach's mouth twitched; his eyes gave away nothing. He ordered the team to take a few laps and then warmups. We were about to run sprints for two hours.

As we turned to go, we heard the half-smile in Coach's voice, "Bott, feel free to use the bathroom," and the laughter erupted.

Practice went fine after that.

-liar-

My father taught Bry, Cousin Rob and me how to play Liar's Dice. Shake up five dice in a cup, overturn the cup onto a table, peek at the results.

Then, slide the cup over to the next player with a claim of what lay hidden underneath. Five twos. Three fours. Four sixes. Believe the person, and you had to roll better than the claim. Call the person a no-good filthy liar, and you had to remove the cup for everyone to see if you were right.

This game turned into a family favorite, especially sliding the dice off the table quickly, then catching them mid-air in the cup before they flew across the room, and the psychological battles between my brother and father. Bry would study our father's facial expressions and body language carefully, ask questions, and then pick apart the answers. He was Vincini ("You've given everything away!") to my father's Wesley from *The Princess Bride*. Sometimes, my father would say he forgot how many fives or sixes he rolled, or he'd change his mind. "Hold on, I meant four threes, not three fours." Sometimes, he wouldn't even look at what he rolled under the cup and just guess, sending his son's mind into somersaults. The contests always left him laughing so much his face hurt.

Me? I knew better. During his time in the service, my father played a lot of poker and won big. Big enough to buy my mother a mink hat, very fancy for a couple of small towners. Also, big enough to pay for a week's vacation—in Hawaii. I never tried to figure out the old man, playing the statistics instead. I was good at math, and he was impossible to read. Who knew what was going on inside that head?

I would be grateful for this ignorance later.

-mill town mascot-

We were not the best team in the league. Leicester had the star player and two very good sidekicks, but we finished a close second. We were still mighty, for we were Woolies.

Millbury is an old Massachusetts mill town of 12,000 mostly middle-class residents spread over 15 square miles of hills. There's a rural feel to the southwestern border and an urban one to the north. For generations, several steel, wire, and textile factories in the center of town backboned the economy with some help from a couple of others in the west. One of the oldest mills, the S&D, built in the mid-1700s looked every bit its age but supplied the Major Leagues with the red stitching for their baseballs.

To this day, jokes fly whenever I tell someone about my high school mascot, not an eagle, a bear, or some other fierce predator. After the initial laughter, the first question: *What is it?* Be wary of Google, our beloved Woolie is certainly not a marijuana cigarette laced with PCP or crack. Other Internet dictionaries ring truer—"having the consistency or covering of wool." In the 80's, pictures and flyers around school depicted an ovine humanoid, smiling, friendly, and cuddly. In more recent artwork, the creature is muscular and busting out of its maroon sweater. So shut the hell up, I say to the gigglers.

Thanks to the varsity promotion, sophomore year as a Woolie was very good to me. I was so gifted at pestering opponents with my defense that I became our starting point guard by mid-season. Coach moved Steven to forward, and just like that we were starters together —from the little Bott-on Garden by the brook to the highest competition in town. Our families would sit sprinkled among the 200 or so spectators, cheering us on, Mr. Bott keeping records of our points in a small black pocket notebook with a mini-golf pencil.

Like most rookies drafted into the NBA, I upgraded my wardrobe: Chess King suspenders and skinny leather ties with Reebok high tops. I loaded up on the hair spray, mousse and gel. Friends were impressed because that's what friends do for each other, but Steven and the other upper classmen on the team also high-fived me or punched me playfully in the school hallways. Girls I didn't know smiled at me. Teachers would tell me that I was mentioned in the newspaper. So many people *noticed!*

As a senior, Steven became more man-like each passing month with his 6'2" wiry frame, dark eyes, prominent cheek bones, and jet-black spiked mullet. He often drew comparisons to the actor Matt Dillon, and girls noticed, including my classmates, including my girlfriend. The charismatic rebel vibe worked for him, and in a surprise to no one, especially Coach Dunham, he was voted Most Talkative in his class.

Steven's relationship with Coach Dunham lived up to expectations. There was the game Steven dove into the stands to save the basketball from going out of bounds and didn't return. Instead, he sat in the bleachers talking to a pretty girl from another town. Coach benched him for the rest of the night. Or there was the time that Steven wasn't paying attention during practice and Coach told him to knock it off or "get outta here and go join a church league!" At the next practice, Steven wore a CYO (Christian Youth Organization) t-shirt, claiming he forgot his practice jersey at home.

It was a great year, and since Steven had helped jump start my varsity career, it was only fitting that I'd help end his.

-sprite-

Our final game that season was at Bartlett High School in the second round of the Massachusetts State Tournament playoffs. We lost by 16 points.

Afterward I sat alone on the bus, ruminating. I'd missed eight of nine shots, all unguarded. With each miss, the conversation in my head grew louder. *Why am I so open? Should I shoot? I should shoot. What if I miss?*

At some point during that lonely ride back to Millbury, Steven slid into the vacant seat next to me and handed me a soda. I took the bottle and drank. The liquid burned slightly, yet tasted sweet with a fruity undertone.

"What is it?" I asked.

"Sprite," he said.

It was definitely not Sprite. I tried handing it back to him, but he waved me off, claiming he had plenty more.

I shrugged and swallowed another mouthful. The soda wasn't half bad when I was ready for the unusual taste.

I stared at the floor for what felt like an hour before eventually mumbling an apology. I never looked at him.

He nudged me with his shoulder and told me to forget about it. We weren't going to beat them anyway, he said. A bald-faced lie. The rest of the team had played great.

I don't remember how he did it but he pulled me out of the funk, joking and teasing as only he could. Before I knew it, we were talking about who knows what and laughing, and even though I had almost single-handedly ruined the last game of his high school career, I felt really good.

As the bus pulled into our high school parking lot, I peered into the empty soda bottle.

"Oh, no! I drank your Sprite!"

He grinned and clapped me on the shoulder, telling me not to worry, and returned to his seat in the back.

Coach Dunham stood up in the front of the bus as it came to a stop and gave a brief announcement on washing and returning our uniforms to him later in the week.

"Way to go, Coach!" I yelled. He was getting his uniforms back. That was wonderful.

He gave me a puzzled look. I smiled at him. It was so easy to smile.

I stumbled off the bus with my teammates, a curiously jovial bunch for having just lost such a big game. Odd, but oh well, I wanted to rap. I grabbed Rem, one of the seniors, as we crossed the parking lot and launched into a stirring rendition of *It's Tricky* by Run DMC. Before I could start up again, he suggested that I continue the concert inside, where more people could enjoy it. A marvelous idea, I agreed.

The hallways echoed with music from a school dance taking place in the gymnasium. I forgot about my encore and rushed to the locker room to put away my gym bag. In the center of the room, two teammates were wrestling, which got me thinking. I jumped onto one of the benches and cheered the action on. Everybody was watching and shouting, except for our big man, Jake.

More Wookie than Woolie, Greg "Jake" Jacobsen, stood at 6'6" tall and maybe 210 pounds. He led our team in points, rebounds, and bone-jarring screens. When a player ran into him, it sounded like a raw steak smacking against a Jersey barrier. Fortunately for us, he was difficult to anger.

I crouched on the bench, watching him stuff his gym bag into his locker and head for the exit. Where did he think he was going?

As I launched onto his back, I wrapped my arms around his shoulders and neck. He flailed and spun, trying to shake me off, but I had locked my fingers together. The team roared and cheered while I held on for dear life. He was so strong! After a few seconds, my grip started to loosen, and Steven and Rem ran up to us to pull me off. Once again, I had the element of surprise on my side. Like a lemur navigating the treetops, I pushed off Jake and leapt onto Rem.

Rem's eyes went wide as I went airborne. Not nearly as big or sturdy as Jake, he bent in half when I landed on him, sending us both crashing to the floor.

"Easy, little man!" he said, laughing as I tried to pin him to the ground.

For the next few chaotic minutes, I tackled whoever happened to be in arm's reach with no real purpose or goal in mind. Eventually I tired and was subdued, and we headed out to the dance.

The following morning, a symphony of steel drums pounded on the inside of my skull. I peeked at the alarm clock from beneath my covers, cursing the light. At breakfast, oatmeal inched down my throat like sunbaked mulch as a single thought slowly formed in my head.

Friggin' Steven.

-posterity-

A few months after the Sprite Game, Steven graduated from high school by the skin of his teeth and began an adult life. He had a car, a nine-to-five at a factory, and a Bott work ethic. He also had a steady girlfriend, Lisa, a high school sophomore from a nearby town—a sweet girl with wavy blond hair, green eyes, and trim figure. I envied my friend's good fortune, but I missed our time together. No more rides to school. No more high school summer leagues. Steven made new friends at the plant, enjoyed his weekends, and played more softball and less basketball.

The varsity boys' basketball team photo from the 1987 Millbury Memorial High School Yearbook showcases a collection of long, thin teenagers topped with oversized, wavy mullets. Coach Dunham is absent from the picture—I'm not sure why—he's in the team photos for all the other yearbooks I've seen.

We sit along two rows of bleachers in the gymnasium. I'm in the back between Scott Turner and Bob Ayotte, giving a half smile. Steven is front and center with the other seniors. Everyone is looking at the camera, except for Grubby, who stares intently off to his left like a retriever that's spotted a rabbit and Steven who's eyed something off to his right and is laughing. He's also up on his toes and leaning to the side, like he's about to jump up and run out of the picture.

I've come across the photo a handful of times over the years and always wondered what was so funny. The moment captures Steven perfectly. Even frozen in time he seems unsettled and in motion. It wasn't until decades later that I discovered the punchline, his hands. Both are resting on his knees, clenched into fists. The lone exception is the middle finger on his right hand, extended straight downward. Clearly intentional. Saved for posterity.

-wall-

Roger Desrosiers was one of my toughest high school teachers, burying his students in reading assignments. I've always been a slow reader. I liked Roger with his dark mustache and small, dark smiling eyes peeking out behind wireframe glasses, but I remember how quickly that smile vanished if you weren't paying attention, how chilly the classroom could turn.

During my junior year, he entered our class into a local competition on the U.S. Constitution, which we surprisingly won. That led to a state competition, which we miraculously won. That led to a national competition, which we didn't come close to winning.

I was never a fan of U.S. History, finding it to be dry and factual and filled with dates and milestones about old white guys wearing wigs.

There were no problems or puzzles to solve or fictional stories to spark the imagination and no apparent connection to me. Everything had been dead and dusty before my great-great grandparents were born.

But the national tournament landed us in D.C. for a week. I fell in love with the city and how the monuments and museums brought the past to life. I could feel Washington's importance standing at the base of the obelisk, growing dizzy as I looked up into the sky until a classmate ruined the moment, "Now, that's an erection!"

I felt Lincoln's burden as he sat tall, overlooking a country tearing itself apart. Winds glided over the Tidal Basin as we circled Jefferson, the sun casting long shadows across his stony visage. We toured the Smithsonian and all its trinkets and wonderful inventions born of our emergent culture. Then, the military monuments: US Marine Corps-Iwo Jima, Korean War Veterans, Tomb of the Unknown Soldier, World War II. We snapped pictures and asked questions. What did he do? Why is this important? What does this represent? We sat on the stairs in front of Lincoln, snacking on pretzels, looking out over the reflecting pool, immersed in the past.

Finally, there was the Vietnam Memorial: a long, black granite wall that descended into the ground, covered in the names of deceased soldiers. I walked alongside and heard a whisper—*down, dark, death*—chilling my blood. Tourists spoke in hushed tones. Even my boisterous classmates had fallen quiet. Nobody voiced the questions we held inside. *What had happened? What was different about Vietnam?* At the center of the monument, the wall stood ten feet high. I considered the thousands of names above me and felt like I was standing in a grave, looking up at the tombstone.

I'd brought a pencil and paper to rub over the names of two soldiers Mr. Bott had asked me to find. Sadly, I've forgotten which two. Maybe one was his younger brother, Pete, a Green Beret who'd been captured and ended up MIA, never to return home. My father didn't ask me to look for any names, but then again, my father never brought up the war.

-leaving-

My last two years of high school blurred. Honors and AP classes and mountains of homework. Study groups, SAT Prep, SAT's, college applications and essays and interviews. The U.S. Constitution competition and a trip to D.C. All the while training, practices, games, and playoffs. Oh yeah, and dating.

I crashed.

The last week of finals during my junior year, I lay across a countertop in chemistry lab, soaking in its coolness. I remember Mrs. Graves asking me if I needed to see the nurse. Two days later, the fever spiked, and I stayed in bed for three weeks with strep throat and mono.

When the dust settled, high school had wrapped up, summer had arrived and I was leaving for Worcester in a few months to attend Worcester Polytechnic Institute, a small technical college—only 20 minutes away, but leaving nonetheless.

Some evenings during that final summer, I'd grab a towel and walk down to the Botts to cool off in the pool. If there were hot dogs and hamburgers on the grill, I was invited to dinner. If there was a cookout or party the following weekend, I'd better be there, and I'd better tell my mother and Bry, too. The first words out of Mrs. Bott mouth whenever she saw me were always "So, how's your mother?"

With college on deck, the Botts made sure I'd be heading off with several nicknames. The classic, *Christopher Robin* had lost steam with Mrs. Bott. In all fairness, it was a mouthful. *Chrissy* or *Chrissy Richards*, also an original, showed no signs of letting up with Joe, John, George, and Mrs. Bott. *Christo,* used only by Fred, the oldest brother, had just started as a response to me teasing him as Fredo. ("I know it was you Fredo. You broke my heart!") And then finally, *Michael,* two years running with all copyrights held exclusively by Mr. Bott.

I chose Michael for my Catholic Confirmation name and asked Mr. Bott to be my sponsor. He made the ceremony memorable by squeezing my shoulder with the force of a hydraulic clamp during the Cardinal's blessing. Most sponsors simply rested their hands on the confirmed.

Not Mr. Bott. As my eyes filled with tears, the stoic clergyman smiled knowingly at me, thinking I'd been overcome with emotion.

Except for *Christopher Robin*, each nickname came with an exclamation point. *Michael! Chrissy Richards!* The guys never just greeted me per standard social conventions, especially Fred and Mr. Bott. I was announced, as if going up on stage. *Ladies and Gentlemen, let's hear it for … Cristo!*

I once heard that we're given two families in life: the one we're born into and the one we choose. In my case, it was the family who named me at birth and the family who nicknamed me in adolescence.

-armpits-

There are maybe five or six boys in the 200-year history of Millbury High School who went on to play basketball at the college level. Brian Blanchard, Brian Parath, Big Jake, and my brother—the ones I remember—were all well over six feet tall. As were all the players on the Worcester Polytechnic Institute, or WPI, men's team my freshmen year of college. So, when I stepped onto the court for our first practice, my face lined up with everyone's armpits, and I wished I'd enjoyed baseball more as a kid.

No worries, though. I'd been a team captain-hot-shot-MVP for a high school squad that lost only one game my senior year. I'd been recruited by WPI's coaches and even won a pair of Avia basketball sneakers at their summer camp. WPI was no Duke or Kentucky. It was a small school for science nerds. I'd be a star.

The long-limbed freaks closed in on me like giant spiders scurrying along a web. Half of them were Jake's height or taller. All of them were faster, even the guy built like an icebox. Apparently, my moves were predictable, and my shot took an eternity. They blocked seemingly open layups, stole routine passes, and engulfed me in a typhoon of hands and elbows.

During tryouts, the coaches shook their heads and stared at their shoes as my mistakes piled up. They excluded me when splitting teams up for scrimmages and assigned me to junior varsity.

Collegiate JV is the equivalent of basketball purgatory. I hadn't been cut, but on the few days we did practice, we shared the court with varsity and mostly stood on the sidelines, watching. At the very end, when varsity had left for the showers, we'd get to scrimmage for a few minutes. A big, fat waste of time. After a week or so of this, I walked into the coach's office and quit.

"I'm sorry to hear that," he said, the surprise in his voice quickly washed over with resignation. I was hardly the first player to jump the JV ship.

"Why don't you take some time and think about it?" he said.

I left his office with no intentions to think about it. I would miss the game and the competition so much it would hurt. But I wouldn't miss the armpits.

-cliff-

Soon after, Joe Bott called.

Which was weird because Joe and I rarely talked and I don't think we ever spoke on the phone. Suddenly, he was interested in college and very interested in the basketball team.

"Don't do that," he sighed after I told him that I quit and would try out again the following year.

Joe didn't understand. On our junior high team, Bobby Ayotte was our best player because he was as tall as a grown-up and could dribble with both hands. The assistant coach was silly and drunk, and Joe gave us rides to games in his White Mustang. Johnny and I laughed all the time, and we wished the season would never end. College basketball was nothing like that.

"Why not?" I asked.

The cliff, Joe warned. There'd be no slow decline, no tapering of skills from competitive sports. Only a plummet back to the abilities of the recreational masses. He spoke of his own experience as a high school player. A year off would be disastrous. I'd never make the varsity he said. I hurried him off the phone, unsettled.

A few days later Steven called out of the blue and wanted to know if I could sneak him and a couple of guys into the gym to play ball on the weekend. Of course, I said. Nobody used the old Alumni gym on weekends, and it would be great to see him.

He brought Bubba and Nelly, two of my favorite Steven friends. Bubba stood tall and heavy and jolly and moved with deceptive quickness. I always thought Big Smooth should have been his nickname—the way he glided and shimmied. Nelly was dry, witty, and unlike Steven in every way. I don't know how they ever became friends. Together, the three made a natural stooges act, the jokes flying without restraint, years of familiarity behind every taunt. I laughed all afternoon. I missed my hometown.

I don't remember the games or outcomes, or if Steven tried to match up with me. I imagine he started strong with his quick hands, good instincts, and heads-up plays. I bet he shined in the first game and maybe even the second. But then there was a third and another 15 minutes of sprinting, his face reddened, breathing labored. Had he gone to the St. Charles Hotel or the Old Time Inn the night before for beers with the softball team? And then came the fourth game and then a fifth and I was still running full-tilt boogie because the team conditioning drills had been insane. But was he falling a step behind the action, making turnovers and missing shots? Had he put in some overtime at the factory that week?

He must have because everything changed that afternoon for me. Joe was right. I couldn't quit. I stood at the precipice peeking over the edge, a witness to Steven's descent. Next in line.

-racquetball-

First, there was catch with my father in the backyard before Little League games. My mother would make her famous American chop suey for dinner, the boiling water kicking up steam and fogging up the kitchen windows while the aroma of oregano, onions and bell peppers filled the house. I'd wait in my little league uniform for my father to get home from work. He'd throw ground balls to me and one-hoppers and impossibly high flies that kissed the clouds and left me staggering to align myself with the return flight. He always threw sidearm. A shoulder injury he said.

Then, there was football behind the apartment with Bry and Cousin Rob, my father at quarterback, throwing posts and slants and deep fades that carried us into the neighbor's yard. The game often ended with the three of us sacking my father and dogpiling on top.

Then, basketball at the court behind the Wilkinsonville Fire Station with the hoops well over ten feet and the pavement covered in sand and dirt after a good rainstorm. My father wore a headband and two wristbands, dribbled fast and low like Bob Cousy, and had the *Wilkinsonville Hook* shot that was un-guardable.

By high school, contact sports were a thing of the past with him in his early-40s and us approaching adult weight and might. So, he and I got into racquetball, and it became our thing. Weekend visits had become intermittent, filled with dates, homework, and playing sports. With life filling up, my father was getting pushed out, but racquetball held us together. The Friday night ritual—an hour of court time at the Main St. YMCA in Worcester, followed by a small cheese pizza or an Italian *grindah* [sub/submarine sandwich] and a Sprite—would continue well into my college years.

Aubrey Greenwood, the manager at the Y, would enthusiastically greet us at the front desk, "Hey, Larry! Hey, Chris!" and then talk with my father about something to do with work. I still didn't really understand my father's occupation at the Social Security Administration. He said he helped people who were injured or sick, but I didn't know how.

In college, I started to win our matches regularly even though he was still the better player. I was okay. I'd kept up with him over the years because I was a quick, conditioned teen, bouncing around the court like a caffeinated squirrel. When I started winning it had more to do with him than me. A weight across his broad shoulders. A weariness behind his cobalt blue eyes. Daily naps, sometimes for hours, and even when awake, sluggish.

Something was wrong. He was sick.

Chronic Fatigue Immunodeficiency Syndrome, he told me one day. CFIDS for short. I'd never heard of it. Neither had any of his doctors, he said.

-forgotten-

One Saturday morning, in the early fall of 1990, Steven picked me up at school and brought me to a gym somewhere north of the city. The sun shone, and the court was small, definitely not a high school, maybe an annex to a church or VFW. We played full-court games, and Johnny Manyak, a good friend of the Botts, stood at the three-point line, laughing. Johnny had an infectious laugh.

That's all I've got. That's the story.

I've tried to recall more, but nothing concrete surfaces—just feelings and extrapolations. I think we were in the town of Holden or maybe West Boylston. Was Bubba there? Probably. Nelly? Maybe. I have a hunch Steven's Uncle David of the sleepy eyes and dark mustache had reserved the gym time. Uncle David played everywhere and with everyone even though he was well into his forties.

Given the approximate date, Steven must've been showing signs. The new friends, the weekend binges, approaching a different kind of cliff. Did I know? I don't think so. Was he wearing sunglasses to cover puffy, bloodshot eyes? Had the beer added a sagging belly to his lean frame?

As for the games, I can't remember a single play. I wish I could find three seconds floating around in my head—a pass to Steven in the corner for a shot or a rebound and we're racing down the court together. Something I could bookend with that summer in his driveway when he helped me make the JV team.

I imagine when he pulled up to my apartment to drop me off afterward, I was thinking about the pile of homework waiting for me on my grey-and-green-flecked laminate desk with the metal trim. I like to think that I asked him to come see my games. I'd be trying out for the varsity team in a month or so. Of course, he'd come, he would've said and smiled.

We would've been in his brand new sporty red Nissan Pathfinder, the first big purchase of his life. After dropping me off, he would've headed north on Fruit St. and taken a right onto Highland, disappearing around the corner. I really liked the Pathfinder. It was a dark red, almost crimson, not a flashy, look-at-me candy apple color.

He'd never make it to any of my games, and I'd never play basketball with him again.

-domino theory-

My father was born in the summer of 1946 and named after his father, Lawrence. A year later, the United States and Russia gave birth to what would be known as the Cold War, with the U.S. hardening its policies against any allies of the Soviet Union to stop the spread of Communism.

When Larry was in third grade at the Sutton Public Elementary School, the small country of Vietnam on the other side of the planet was split into two during the Geneva Convention, with the North becoming a communist regime. Supposedly, Larry was a big, heavy kid, and fights were common. "Sometimes I was the bully, sometimes I was the target," he once told me.

By the time Larry had moved up to the Middle School, a communist insurgency had begun in South Vietnam. The United States believed that the communists in the South (Viet Cong) were supported and guided by the Soviet Union and China. Policymakers warned that if South Vietnam fell to Communism, neighboring countries would inevitably follow suit, one after another like a row of dominoes.

When Larry was 13 and taking long walks with his collie in the acres of forest behind his home, North Vietnam was openly sending ammo and men into the South. A year later a team of government advisers recommended the U.S. increase military, economic, and technical aid to South Vietnam to confront the growing Viet Cong threat. President John F. Kennedy increased the aid but stopped short of committing to a large-scale military intervention.

During the fall of his senior year of high school, Larry broke his thumb at the beginning of what should have been a very promising basketball season; he was being recruited by the coach of Assumption College in Worcester. Weeks later, South Vietnamese President Ngo Dinh Diem was killed during a military coup and President Kennedy was assassinated in Dallas. The ensuing political instability in South Vietnam persuaded Kennedy's successor, Lyndon B. Johnson, to further increase U.S. military and economic support.

The following summer, Larry graduated from high school and North Vietnamese boats attacked two U.S. destroyers in the Gulf of Tonkin. President Johnson ordered the retaliatory bombing of military targets in North Vietnam, and Congress passed the Gulf of Tonkin Resolution, giving the President broad war-making powers.

-gravity-

I sat at the grey-and-green-flecked laminate desk with metal trim and legs when the phone rang and the world flipped. My mother's voice was hoarse and tight. She paused, stammered, then heaved a 100-pound sigh.

The desk was unsightly but functional, and it came with the apartment; a second-floor unit of a three decker with a narrow, uneven driveway that ended in a tilted garage. There were two windows in the room. One provided a view of the vinyl siding of the adjacent building to the south; the other looked out over a rusting chain link fence at an unused back porch.

On the desk lay a pile of homework; physics. We were studying objects moving, bumping, bouncing... and falling.

There was an accident at work, she said.

Falls happen. I was maybe two when I slipped through the railing at the top of the stairs in our duplex, dropped like a stone, and smacked my head on the floor. Mild concussion and fractured skull, but okay. A couple years later a kid pushed me off a ten-foot-high deck. Another mild concussion; another fractured skull. Fine in a week or so. Falls happen.

I'd even seen Steven fall. We were kids, 12-ish. One lazy summer afternoon, we came across a mustard metal origami the size of an elephant parked alongside the road. A digger of some sort had been abandoned by the local Public Works crew for the day. Being the curious types, we hopped off our bikes and explored.

While I searched along the massive tires and underbelly, Steven climbed up to the driver's cabin. Moments later the squeak of a sneaker and a startled yelp rang out through the quiet neighborhood.

I rushed over to find him lying on his back in the road, clutching at a long crimson slice along his leg, trying to squeeze it shut. The blood spread in little streams beneath his fingers, and his eyes grew wide. My heart hammered away as I turned and sprinted home for help.

Several stitches and a few weeks later we were back on our bikes.

Steven had fallen at the factory, my mother said. He was at the hospital. No, we couldn't go visit, not yet. Most details were unknown or withheld. But what kind of information clammed up the family herald? I backed off the questions, the answers lurking like low growls in a dark cave. She promised to call back when she knew more.

The air in the room grew cold and thin. Knots filled my stomach like softballs.

My homework lay before me.

If a 2-pound object sits at a height of 20 feet, what is the potential energy? Steven was hurt. *If a 5-pound object drops 30 feet, what's the velocity when it hits the ground?* At the hospital. *How long does it take for the object to hit the ground?* What was she not saying?

I stared at the physics book. Nothing happened, everything happened, and worlds merged. They say mathematics is the language of the universe.

If a 200-pound man falls ... say 20 feet ... what is the impact force when he hits the ground?

I wanted the math to tell me that he was okay.

Impact force. Distance.

Gravitational Constant.

How high did he fall from?

In the end, it didn't matter.

Whether he fell ten feet or 20, the resulting forces lit up my tiny TI calculator display with more zeros than I could fathom, sucking the remaining air out of the room.

I'd miscalculated, of course. I ran through the equations again. Same answer. Again. Same answer. The bedroom walls crept closer.

I sat and stared at the numbers until I needed to stand, and then stood until I needed to pace, and then it all became too much. Shoving a pair of sneakers, a jump rope and a basketball into a gym bag, I slipped out the back door, undetected by my roommates. Nobody used the back entrance—a small creepy, unlit hallway and sloped stairs with a strange, earthy scent.

I imagine the air that evening was brisk and carried the scent of gasoline. Moving from a suburb to the city, I always noticed the gas. Cut grass, grilled streaks, wet leaves, and wood burning stoves had been replaced with diesel 365 days a year. Maybe a whiff of onions or garlic near a restaurant. Or the sewer after a rainstorm. But always fuel.

Alumni was a dusty, old, crappy gym. I loved it. A running track on the second floor overhung the corners of the basketball court, and the night sky filled two large windows at opposite ends of the building high above. Tired halogen lamps cast a dim, amber glow that barely reached the empty bleachers, leaving deep shadows.

The jump rope whirred and snapped against the floor, and my spinning mind slowly receded into the gloom. A few minutes later, I dropped the rope and picked up the basketball.

For the walk home, endorphins had taken the wheel. No more intestinal twists. No more tightness across my chest. *Everything was gonna be alright!* Steven had just injured his hand or wrist, bracing himself when he fell. Maybe a nasty compound fracture. Probably a surgery. I'd visit once he was up for company and remind him of the DPW digger and other childhood misadventures, and we'd laugh.

The next time my mother called, I was at my desk, lost in physics again. Choking back tears, she told me that Steven had fallen 50 feet and landed on his head, that he was in a coma, and that his family had been summoned to the hospital to say their goodbyes.

-visibility-

In June 1959, an experiment led by Dr. James Brown at Camp Drum, New York, demonstrated the long-term effectiveness of aerially dispensed herbicides in improving visibility for military operations. An improvised helicopter spray system delivered a 1:1 mixture of 2,4-dichlorophenoxyacetic acid and 2,4,5-trichlorophenoxyacetic acid over a 4-square-mile area at a quantity of one-half gallon per acre.

Evaluation of the effectiveness of the defoliants on vegetation was made one year later and again in October 1962. For the first evaluation, no signs of regrowth were observed in the sprayed area. Upon reexamination in 1962, it was noted that all the maple trees, which had been predominant in the area, appeared to be dead. In general, trees throughout the area had been killed, and visibility had been improved nearly 100%.

The United States began the use of weapons-grade herbicides in Vietnam later that year (Buckingham, 1982).

-rule number three-

During childhood, one of the most popular games among our pack of boys was *Guns!*. It required a minimum of two kids and could be played by anyone.

The first rule of *Guns!* is this: everyone needed a gun. This could be a toy gun, a water pistol, a stick, or even a small household appliance if your mother wasn't looking.

The second rule of *Guns!* is this: Players are divided into two teams—hiders and seekers. Seekers allowed hiders somewhere between 20 seconds and ten minutes to ready their ambushes in the woods, house, yard, or whatever area defined the playing field. Then seekers went seeking, and amazing battles ensued. If your gun didn't come with its own sound effects, you made the appropriate noises: high-pitched squeals or whistles for lasers; deep, wet bass-vibrations for bazookas and rocket launchers; choppy barking for machine guns; and the tried and true 'BANG!' for revolvers.

The third rule of *Guns!*: when you shot someone they *died*, eliminated from the game. The game ended once all the players from a team had been shot.

Ingenious in its simplicity and flexibility, the game suited any number of children and could be played anywhere for any period of time. It all worked. There were also infinite variations, such as *Cops and Robbers*,

Humans and Aliens, Zombies and Werewolves. The variations affected the weapons of choice and sound effects, but the basic rules remained the same.

The lone weakness of *Guns!* lay in the subjectivity of rule number three. Since no concrete evidence existed that a player had been shot by a toy gun with an imaginary bullet or laser, anyone could dispute the point so that they could keep playing. As a result, action-packed battles sometimes became action-packed arguments. Over the years, the game would undergo dramatic upgrades with the inventions of Nerf, laser tag, and even paintball, but before these versions came along, everything ran on imagination and the honor system.

In all the years we played *Guns!*, Steven never died. Not once.

Oh, there were many close calls, such as the time we had him surrounded in a tree, five against one, or the time John snuck up behind him and shot him in the back point blank. And who could forget the afternoon I caught him kissing a girl behind a row of bushes. Steven claimed misses on all accounts, escaping his predicaments.

Yes, he exploited the subjectivity of rule three to his advantage and beyond any credibility. Bullets only hit you if you agreed that they hit you, and you didn't die unless you agreed to be dead. So, he never agreed. Of course, we proceeded as if he had been eliminated, but in his mind, we never got him.

So, as the hours after his accident turned into days and days turned into weeks and Steven did not die as doctors had predicted, I thought about our time back in the neighborhood, his "perfect record" still intact.

You missed me. I'm not dead.

-concern-

In January of 1966, 29 scientists banded together to protest the U.S. policy on the use of herbicides in Vietnam and demand their complete abolition. They requested that President Johnson begin discussions with the allies on adherence to a ban.

Even if it can be shown that the chemicals are not toxic to man, such tactics are barbarous because they are indiscriminate; they represent an attack on the entire population of the region where the crops are destroyed, combatants and non-combatants alike. [This is] ... a precedent for the use of similar but even more dangerous chemical agents against our allies and ourselves. (Dux and Young, 1994)

When the government failed to act, the Council on the American Association for the Advancement of Science (AAAS) sent a letter to the Secretary of Defense, Robert McNamara, calling for studies of the short- and long-term consequences of the massive use of herbicides in Vietnam.

-vanishing act-

Right before his accident, Steven had been strolling along an outdoor catwalk with a box of filters in his hand. He didn't normally work in the factory's filter house, but business had slowed during the summer and talk of furloughs circulated throughout the plant. Instead of taking a pay cut he took on hours cleaning smokestacks—a shit job, but it kept him working until business picked up again.

His friend, Vito, waited at the end of the catwalk and asked Steven for the time. Instead of answering, Steven vanished like a magic trick, the floor giving way beneath him. Vito sprinted off into the factory, shouting for help, while Steven lay face down, unconscious on the pavement below.

The surgeons at the UMass Medical Center in Worcester drained fluid from his brain and rebuilt his skull (a closed head wound, they called it). They performed a gastrostomy to provide him with a feeding tube and a tracheostomy so he could breathe. They used plates and screws and bone from his hip to reconstruct his right hand, wrist, and forearm, a surgery that took nine hours alone. In short, they saved his life, though he remained unconscious.

Three weeks later, doctors noticed during a routine chest x-ray that he also had a broken neck, leaving everyone baffled as to how the injury had been missed with the endless battery of x-rays, CT scans, and MRIs. Instead of another surgery, doctors mounted a halo brace onto his head and shoulders to stabilize his neck so that the bone could reset.

Two weeks after that Steven woke up from his coma.

-letter-

Larry was mid-way through his junior year of college at UMass, pulling a D+ average in "ping pong, pool, and poker" when a letter arrived from the U.S. Armed Forces; in two weeks he was going to be drafted for two years of service.

Vietnam was escalating. The anti-war movement was growing. Young men were fleeing the country.

The letter also stated that Larry had the option of enlisting. Enlistment meant three years of service in the army or four years in the air force or navy. But he could choose the branch and the type of service. If he was drafted, he had no say in where he would go. It could be the front lines.

Larry's father had served in World War II as a shipman on the USS Macomb, a high-speed minesweeper destroyer that cleared 213 mines, sunk the German submarine U616 and shot down seven Japanese aircraft. They'd hunted the Bismarck in the Atlantic and survived a kamikaze strike in the Pacific. They were going to be one of the lead boats for what was to be the final invasion of Tokyo.

"We were all going to die," my grandfather once told me as he lit a cigarette. "Japan was never going to surrender."

And then Nagasaki and Hiroshima happened, and the Tokyo invasion didn't. My father, my brother, and I—we were alive because of the atom bomb. Eerie.

My grandfather told Larry to join the Navy and stay out of the jungles. Vietnam was a jungle war. But he'd also shared with his son some of his war stories of being out at sea. Larry was terrified of drowning.

Two hundred American soldiers had died in 1964, 2,000 in 1965, 6,000 in 1966. It was 1967. Larry was 21 years old.

-visit-

The trip to the Greenery Rehabilitation Center in Brighton was a 45-minute straight shot east on the Mass Turnpike. My mother's boyfriend, Jack, drove us as I stared out the window at the bare trees whirring by in an endless undulating mass. I don't remember the sky, but most December skies above Massachusetts are gray.

Dread had settled across my chest like a thunder cloud, but I'd hoped to cheer Steven up a little and maybe get a laugh—or at least a smile. Then, I walked into his room.

"Holy shit!" I whispered.

I slowed my pace, allowing my mother and Jack to move ahead of me. I'd been studying Elie Wiesel for a philosophy paper at school and the project had been haunting me, so much that when I first saw Steven, I thought of the concentration camp prisoners. This had to be an exaggeration. He couldn't have been anything like those photos, the Auschwitz undead with their hollow, lifeless eyes and skeletal bodies.

Steven lay on his back. Encircling his head, a ring of thick black metal bars was held in place with screws piercing the skull like plywood. Dark stains filled the threads closest to the skin, and a hose punctured his throat with more stains. The hose connected to a machine near his bed that whooshed softly. The clavicles and condyles in his shoulders shone against the skin. His legs, a fraction of their original size, kicked and moved and kicked, running a race to nowhere. They tried to kick off the sheet, but it was held in place by his right hand, which sat motionless and enormous, mummified in bandages.

I stepped up to the bed, opposite Jack and my mother. Steven's eyes had locked in on them even though he couldn't turn his head.

"Hi, honey!" my mother said, "It's so-o-o good to see you!" She spoke loudly, causing my brain to erupt with questions. *Can he hear us? Does he understand what we're saying? Does he remember us? Can he communicate? How will we know?* There was no one else in the room to ask.

My mother placed her hands on his shoulder and gave it a squeeze; a hug was not an option with the halo brace and air tube. His eyes remained on her, but nothing in his gaze changed.

"Did you see who came to visit you?" she asked and pointed across the bed at me. His eyes shifted to find me. *Is he following her finger? Does he understand her words?*

"Hey, Steven," I said, and our eyes locked. I searched his gaze for something, anything, but all I saw was an attempt at concentration.

Jack began telling a story to lighten the mood, but Steven didn't turn back to the sound of his voice. He continued to stare at me, and that's when I knew none of this was real. It was some sort of nightmare because if he chose to look at me, he wasn't just following sounds, some part of him was still in there, trapped and he was trying to say *Get me out of here, man!* I just needed to wake up and go to school or wherever, so I looked down at the floor because it had started to shift and sway. Then, I looked up at Jack, who was still telling his story. I tried to smile, but no smile would come. Then, I looked back to Steven, and his eyes were still on me with that focus, staring through me. My mother noticed and said something like, *Look at that, he remembers you!* That's when I rushed out of the room as tears slipped down my cheeks.

The restroom had dull yellow tile. The water was cold on my face. I took long, slow breaths.

When I returned, Jack and my mother were talking to Mrs. Bott, whose eyes told a story I've never forgotten. She gave me a hug and asked about school. I shrugged.

"Steven, did you see who's here?" she said. She spoke at a normal volume, and I wondered if his hearing was fine. I don't remember asking her, but I remember standing just out of Steven's line of sight, a coward.

Until he could eat and breathe on his own, she said, Steven's home would be the Greenery. To help with the breathing, doctors had placed a stent, or maybe a filter, she wasn't sure which, below his heart, The staff was keeping a close watch for any swelling in his head, which would require an immediate trip to the operating room for another procedure. His right wrist had suffered the most damage and the doctors were just happy that he had something that resembled a hand. Most believed that it would never function. His hip was healing nicely. It had survived the fall unscathed so doctors had harvested several pieces of bone from it to help repair his hand. Steven had undergone five surgeries and was on a hearty regimen of anticoagulants, anti-inflammatories, antibiotics, and painkillers.

Eight weeks! I thought. The halo, the tubes, the kicking legs—*this* was what eight weeks of healing and recovery had manifested. What was the first week like? What had happened during the second?

After an almost silent ride back home, I walked into an empty apartment, a rarity with three roommates. In my room, on the green-grey desk, a small pile of textbooks sat waiting for me and the realization hit me like a dump truck: I had to study. I had class the next day. I had assignments, labs, quizzes, practices, and games. Nothing would stop for Steven.

I wondered why there weren't equations for emotional energy. Nervous breakdowns, panic attacks, broken hearts. A language existed, but no mathematics, no formulas. Yet, that grief and loss churned and pulled inside just as strongly as gravity held my feet to the floor, and I hated the stillness in the apartment. I wanted there to be noise. I wanted shouting! Yelling! Explosions! Something to align with the cacophony inside.

But there was nothing. Only indifferent silence.

-qualified scientists-

Shortly after Larry enlisted in the Army, a petition signed by more than 5,000 scientists, including 17 Nobel laureates, was delivered to President Johnson, requesting that he end the use of herbicides in Vietnam.

A Department of Defense official, responding to the growing criticism of the military use of herbicides, stated that "qualified scientists, both inside and outside the government, and in the governments of other nations, have judged that seriously adverse consequences will not occur. Unless we had confidence in these judgements, we would not continue to employ these materials" (Dux and Young, 1994).

-questions-

My brother and I dedicated our basketball seasons to our friend, writing *SB* in permanent marker on the back of our sneakers. Bry, a junior, starred for the Millbury varsity team, and I served as the sixth man for the WPI varsity. Our lives were moving on. Steven's lay trapped in a hospital bed.

The holiday semester break presented another opportunity for a visit, but this time I went alone. Steven hosted a small crowd: his parents and John. Mr. and Mrs. Bott continued to limit visitations to family and a few close friends, with Coach Dunham being the lone exception. Coach had brought Steven a basketball, a Millbury High School warm-up jacket, and a pair of sweatpants. The Woolies back home were thinking of him.

When I entered the room, John greeted me with a bear hug. It'd been a long time. Like his father, uncles, and older brothers, Fred and George, he'd enlisted in the Army after high school and carried himself as military men do—clean-shaven, ramrod posture, and air of subtle confidence. We'd written letters once or twice, but I received most updates from Mrs. Bott. She'd mentioned his transfers to North Carolina and Washington on different occasions.

I imagine there was banter. Was *Stripes* an accurate portrayal of the military? Did college life resemble *Animal House*? With John and his parents there, I remember feeling steadier than during my first visit.

"Hey, Steven, look who's here," Mrs. Bott said as she led me to this bed.

Steven had gained some weight, and his kicking legs kicked more slowly. An overall improvement, though I still found it difficult to look at him with the black halo bolted into his head and breathing tube plugged into his neck.

"Do you know this guy?" She pointed across the bed at me.

Steven's gaze gradually worked its way over. No smile. No emotion. Maybe a flicker of recognition.

"Hey, Steven," I tried to sound chipper.

Steven's left hand moved slowly up from the bed. As the seconds ticked by, his fingers circled and closed into a fist and his thumb extended up toward the ceiling.

Mrs. Bott grinned. "That's right. I knew you'd remember him."

Mr. and Mrs. Bott had worked out a thumbs up-thumbs down language with Steven to communicate yes-like versus no-dislike for basic questions. *Are you cold? Are you in pain? Do you know who I am?*

John soon replaced Mrs. Bott on the other side of the bed, gazing down at his brother. When Steven had fallen, John was in the midst of a month-long series of field exercises in the middle of nowhere, four hours from Fort Lewis, Washington. It took a call from Mrs. Bott to the American Red Cross, who reached out to the commanding officer at the base, who immediately sent a driver to fetch John. The officer instructed John to call home because "your brother has been in an accident," not realizing that John had four brothers. By the time he finally spoke with George, who was waiting at home for the call, Steven had survived the surgeries and slipped into a coma. There was nothing to be done.

"I wish I could take his place," John said softly, his eyes never leaving Steven. "You know, so he wouldn't have to go through this."

The soldier, protecting his family.

No way in hell, I thought, looking at the halo, the tubes, and what was left of Steven. Somewhere inside, though, something else—guilt, overdeveloped, twisted Catholic guilt maybe—was taking shape. Steven's life had been blown up while we laughed and joked, enjoying the holidays. The pangs ran deep after that first visit to the Greenery when I realized I had to study for class the next day. No time to stop and reflect. If I wanted to keep up with school, with the team, I couldn't stop, either. I'd left him behind.

I asked John about the brain injury and memories and personality, talking over Steven as if he were a bowl of fruit. Would the brain heal? What would he remember? What did the doctors know? Not much, John said, until Steven could communicate.

Soon after, maybe even the drive home, I tried to imagine a friendship with no memories or a new personality and limited cognition. All of it or any of it meant starting over, which wouldn't happen. Our childhood, like most childhoods, was unfiltered and powerful in the bonds formed, and that time together had passed. Fifteen years would be wiped clean. Some version of Steven may have survived, but not the friend who had been an older brother to me through a difficult adolescence. That friend would be gone forever.

-ominous-

Larry waited at the airport in Springfield with a hundred other recruits from Massachusetts, New York and Connecticut. They sat in plain clothes, lounging, reading the newspaper or magazines, chatting about their families. Basic training awaited.

The transport soon landed, and they could see it taxiing up to the gate. The waiting area fell silent.

When the plane stopped, one of the propellers fell off the wing onto the tarmac.

"You never saw so many guys go apeshit at once," my father had said.

-favorite son-

One day, Fred Bott, the oldest of the five brothers stopped in on his lunch break to visit Steven. A state police trooper who often patrolled the Mass Pike, Fred had worked out his schedule so that he spent more time covering the stretch of road near the Greenery.

Everyone loved friendly, strong, and good-natured "Saint Fred." So much, in fact, that John and George once asked their parents which of their sons was their favorite.

Mrs. Bott gave her best impression of a diplomat. "I love all my children equally. Each one is a blessing."

"Fred," Mr. Bott said without looking up from his newspaper.

John and George laughed. Fred was their favorite, too.

Given his golden retriever-like disposition, I had a hard time picturing Fred as a police officer, chasing down and arresting violent criminals. He certainly had the physical strength to handle himself, built like a linebacker, but I'd never witnessed anything but kind and pleasant Fred, who made you feel completely at ease when talking with him. I suspected that he played the good cop during interrogations.

For his visit that day to the Greenery, I picture him greeting his parents and approaching Steven's bed with the same uncertainty everyone felt. *What did Steven remember? What was he thinking?* Fred would have rested a hand on Steven's shoulder and squeezed gently, and Steven's eyes would have held him in that unreadable gaze. Fred would say hello and tell Steven that he loved him. He might have wondered if the police uniform made it harder for Steven to recognize him. All of this is conjecture on my part.

What I do know, as told to me and everyone else who knew the Botts, is that Steven greeted his brother by lifting his left hand off the bed and making a fist with his thumb extended upwards. Then the thumb pulled back into the fist, and his wrist rotated so that his closed palm faced the ceiling. With great precision and control, the middle finger slowly extended like a crane, aimed directly at Fred.

And with that we all knew: Steven was still Steven.

-assignment-

In April 1967, Martin Luther King Jr. spoke at Riverside Church in New York about the war, stating that

> ...somehow this madness must cease. We must stop now. I speak as a child of God and brother to the suffering poor of Vietnam. I speak for those whose land is being laid waste, whose homes are being destroyed, whose culture is being subverted. I speak for the poor of America who are paying the double price of smashed hopes at home and death and corruption in Vietnam. I speak as a

citizen of the world, for the world as it stands aghast at the path we have taken. I speak as an American to the leaders of my own nation. The great initiative in this war is ours. The initiative to stop it must be ours. (American Rhetoric, 2010)

A few days later, Muhammad Ali refused to go to war at the Armed Forces Induction Center in Houston, having stated previously that he had "no quarrel with the Viet Cong" (History.com, November 16, 2009). He would be stripped of his boxing title and sentenced to five years in prison.

Around that same time, Larry had finished up basic training and was working at Fort Lewis in Tacoma, Washington when he received his assignment. He was to be stationed in a support position in Heidelberg Germany, known for its great fencing tradition. He loved fencing!

Years later (prior to the divorce), my father took a fencing class. He'd come home at night, energized and telling stories. His teacher had nicknamed him "The Wall" because he never moved during the exercises. He'd stand in one position and defend and counterattack. My father had said it worked against the other students, but the teacher would pick him apart. He taught Bry and me how to parry. Parry one! he'd yell, and we'd move our wiffle bats to the left. Parry two! and we'd move them to the right. Parry three! and we'd turn our wrist and swipe away a low attack. I loved parry three.

Larry called Marie and told her the good news. He would be working in Washington before being transferred to Germany. She could come with him. She moved to Tacoma, and they made plans to get married later that year.

-charmed life-

A few weeks after his single finger salute, Steven suffered a series of severe headaches, leading doctors to suspect that his broken neck hadn't set properly. After a particularly miserable night, he was transferred back to

UMass where surgeons, harvesting bone from his hip yet again, fused his neck and reinstalled the halo.

Unsatisfied with the care at The Greenery, Mr. and Mrs. Bott fought to have their son admitted to Fairlawn Rehabilitation Hospital, 15 minutes away in Worcester. Fairlawn typically didn't take patients with breathing and feeding tubes, but for Steven they made an exception.

The staff set two goals: get Steven functional for daily living activities and then get him home. He needed strength. He needed to learn to swallow and speak. And he needed to learn how to move, grab, twist, push, pull, and lift all sorts of objects with his non-dominant hand and damaged brain.

For my first visit to Fairlawn, family and friends had packed his room. Since Worcester was close by, a steady current of people from Millbury and from his job visited often. He still couldn't talk, but he had his thumbs up-thumbs down signals, his middle finger for Fred, and the sign language signal for "I love you," taught to him by Mr. Bott. The therapists also provided a picture board of day-to-day items—a toilet, a television, a shower—that he could point to if he needed something.

The next time I visited, Mrs. Bott was the only person there. The head of Steven's bed had been raised so that he sat almost straight up, and the feeding and breathing tubes had been removed. Even with the halo still in place, the absence of the trach hose made a world of difference.

I don't remember what we "talked" about, but we must have covered *Steven Bott Day*. For their league championship game against Uxbridge, Coach Dunham and Terry had organized a special event, dedicating the game to Steven. In a show of unity, the team hung banners around the gymnasium and presented Mr. and Mrs. Bott with a Millbury basketball warm-up suit for their son. Millbury won the game in front of a sold-out crowd thanks, in part, to Bry, who had been recently nicknamed, "BIG" by his math teacher. After the game, Mr. Bott took pictures with the team to show Steven.

The support didn't surprise me as the Botts were well-liked throughout the community. With five active boys, Mr. and Mrs. Bott had formed friendships with several different groups of parents. Throw in Mr. Bott,

Fred, George, and John's military service, Joe's involvement with youth coaching, and Saint Fred, everyone's favorite resident, it was easy to see why the town rallied behind them.

When we ran out of things to say about Steven Bott Day, I probably asked the standard hospital visit yes-no questions—*How's the food? Is this better than the Greenery? How's the therapy?*—with him giving me a thumbs-up or thumbs-down and Mrs. Bott elaborating where she could. I'm sure he appreciated the company, but I remember the guilt clinging to me. My life hummed along: an A on the Wiesel paper, scoring ten points in a varsity game, a new girlfriend—all of this happening while he re-learned how to eat solid food.

After that visit, I made sure conversations focused less on me and more on general topics: the weather, the Celtics, our families. Even Bry's basketball season felt okay to talk about. Just not mine. I lived a charmed life, and he sure as hell wasn't going to hear about it.

A few weeks later I received a greeting card from my mother.

Hi Chris,

> I've tried calling a couple of times, but I guess my messages don't get through. Sorry I didn't make your game Wednesday—it would have been close to midnight before I got home. Jack left for Florida yesterday. He'll be gone until Monday. I went to see Steven yesterday afternoon after my doctor's appointment. He motioned to his father that he wanted to show me something. Chris—he's starting to move his foot on the right side all the way up to the ankle! Not only that, he lifted himself up with that bar that hangs over the bed! Mr. Bott came in Monday and found him sideways in the bed and asked him how he got that way, and Steven pointed to himself. Mr. Bott said, "Yeah, right!" But sure enough, he had done it himself. He's really coming along. Ginger sends a lick and a tail wag. I'll see you at the game Saturday.

Let me know if you'd like to go out after the game for something to eat.

Love, Mom

-wicked witch-

Larry and Marie were married in November of 1967. Supposedly during the rehearsal, the priest told Larry that he gave it six months before a divorce. Larry sputtered and cursed his way through the ceremony, wanting to clock the guy.

A few days later, packed and ready to go on a budget honeymoon to northern Vermont, the young couple opened their wedding gifts and realized they had enough money to take a real trip. So, they booked a last-minute flight to the Bahamas with sweaters, boots, gloves and wool socks in tow.

"We each had maybe one outfit to wear the entire week," my mother said.

It wasn't until after the honeymoon that Larry remembered that he wasn't supposed to leave the states and had technically gone AWOL. No one ever brought it up.

"Did you and Mom ever end up going to Germany?" I asked.

"No, my orders were changed," my father said. "They sent me to Vietnam instead."

"Why?"

"Because the Army has two departments," he said. "The Fairy Godmother and the Wicked Witch. You never knew which one was in charge."

Supposedly Marie had a solid political connection through her father. She had offered to make a phone call to try to keep Larry away from the war. Larry said that wouldn't feel right. He said he would make the best of it and consider the tour an adventure.

-homeward bound-

Weeks passed. Winter warmed into spring. Steven's waking hours became a constant stream of therapies. I paid one last visit.

The top half of the bed had been raised so he could watch the small television in his room, and I think he was watching a baseball game. He looked worlds better than at The Greenery, the scars on his throat and head less raw, and the hair on the scalp above his left ear filling in. He'd also put on a healthy amount of weight. The bandages had been removed from his right hand although it remained in a splint, and he sat fully upright in bed. A far cry from the Steven prior to the accident, but at least he had some *life* to him.

Mr. Bott had left to run a quick errand. I sat down in a soft vinyl chair next to the bed. It was just the two of us, and I asked him about heading home.

And he answered. "Yhesss," and then, "Ihna fhew wheeks." The words oozed out, slow and slurred. Messy, wonderful words.

For the first time in months, we talked. He answered most questions with a nod or head shake. When he did speak, his breath was pronounced. A deliberate inhale, fighting to fill his lungs. Then an unsteady exhale that added h's. Sometimes, he ran out of air too quickly and couldn't finish the sentence and had to refill. Then, a light cough and clearing of the throat, and I wondered what that was about. So much effort.

At one point during the conversation there was an attempt at a grin (I might have asked about a cute therapist.), the left side of his mouth reaching for the ear, while the right side nudged a little. Later, there was a half-shrug, the left shoulder pulling up, the right never receiving the signal. At that time, I didn't know about the spasticity in the right leg or the nerve block injection needed to stop the shaking. His personality was intact, the brain was okay, he'd be okay, I told myself repeatedly. Prayers were being answered. Hopes realized. Just a matter of time.

But that right side. *Wake up!* I wanted to shout at it. *Move!*

I thought of stroke patients and hemiparesis and the quiet moments in our visit grew heavy. When Mr. Bott returned, I hurried off, trying to ignore the awful questions clattering in my head.

-welcome-

On January 30, 1968, the Viet Cong and North Vietnamese launched a massive offensive across South Vietnam. It was the first day of Tet—Vietnam's lunar new year and most important holiday. Many South Vietnamese soldiers, expecting an unofficial truce, had gone home. The Viet Cong, known for guerrilla tactics, had never launched an offensive on this scale; consequently, U.S. and South Vietnamese forces were caught completely by surprise.

On the first day of the attack, tens of thousands of Viet Cong soldiers, supported by North Vietnamese forces, overran the five largest cities of South Vietnam, scores of smaller cities and towns, and several U.S. and South Vietnamese bases. The Viet Cong struck at Saigon, South Vietnam's capital, and even held the U.S. embassy for several hours. The action was caught by U.S. television news crews, which also recorded the brutal impromptu street execution of a Viet Cong rebel by a South Vietnamese military official.

In early February, with the conflict at its bloodiest, Larry was on route to Vietnam. The long flight from Washington in the jumbo jetliner with stewardesses, reclining seats, meals, and beverages had been quiet and uneventful. As the plane approached Saigon, they had to delay their descent because an F-100 jet fighter was coming in fast for an emergency landing. Circling above the airport, the men could see the small aircraft out the cabin windows, a trail of black smoke following it.

When they finally landed, Larry and the other soldiers were hurried past the wreckage of the fighter jet and crew working desperately to extinguish the flames and onto a bus to be transported to the military base. The inside of the bus was lined with metal caging all along the windows.

"Why the caging?" someone had asked.

"The VC like to throw grenades through the windows of any buses that might be carrying Americans," the driver said. "We think the caging might help protect them."

Larry boarded the bus.

-golden years-

The Fairlawn Rehabilitation Hospital discharged Steven in late May of 1991, seven months after the accident and a few weeks shy of his 22nd birthday. I don't remember how long I waited before that first visit, but I remember the chair lift at the top of the staircase; the Bott's lived in a 3-story home.

Steven sat in his wheelchair in the center of the living room, facing the TV. Mr. Bott sat in a recliner to the left of his son, and they wore radiant smiles. Steven basking in being home. Mr. Bott basking in having his son home.

I'm sure we talked local sports, family, and school. I'm sure Mr. Bott asked me about college, and I told him about different classes and assignments. He'd always scratch his head, impressed. Or he'd ask me about summer jobs, and I'd tell him about mowing lawns and stocking shelves. He'd nod, impressed. For some reason it was okay to talk about my life with everyone back home.

Steven kept up with the conversations, smiling and chuckling when appropriate, interjecting jokes, confirming what I'd witnessed at Fairlawn; his memory and cognition were comparable, if not equal, to pre-accident. With that, a sense of optimism crept into the room, waiting in the corner, uncertain.

Later, however, when I stepped back and thought about the ripples spreading out from the accident, the world weighed ten tons. Mr. and Mrs. Bott had just finished raising their five boys, spanning three decades from Fred's birth to John's high school graduation. They'd provided, protected, and loved their children, all while Mr. Bott was on the back end of 20 years of military service, including a tour of duty in the Korean War and two in Vietnam. The golden years should have been on the horizon for them: grandchildren, retirement, and travel. Instead, Steven was home and needed around-the-clock care: showering, dressing, eating, cleaning, and transporting. Therapy and exercise would take him only so far, the doctors had said. He'd always need help each day, every day. A long slow grind of a life was taking form.

When I thought about all of the changes in the Bott house—the shifted furniture, the chair lift, the wheelchair—the most disconcerting remained invisible. Not an addition or modification, but a continued absence. That no one in the family had mentioned anything led me to assume the worst.

Soon after, I caught up with John on the phone and asked the question I didn't want to ask, and he told me exactly what I didn't want to hear. Lisa, Steven's girlfriend, was gone.

-karma-

My father told me that he believed in karma, explaining the rules to me. I didn't know what to make of it, but how could I not think of Steven, who likely woke every morning to a concert of aches and shooting pains, calling for his mother or father to help him out of bed and into his wheelchair and then help him from the wheelchair to the toilet. Then, there was maybe a shower and getting dressed and onto breakfast, and God, it was probably 9:00 or 10:00 in the morning. Then came the therapy appointments, getting Steven down the stairs and into the Pathfinder somehow, I didn't want to think about the rest of his day.

If karma was real, Steven would've gotten an arthritic middle finger or maybe male pattern baldness for his countless adolescent transgressions. Not this. This was something else.

-helplessness-

I had originally selected civil engineering as a major in college because I understood bridges and dams and buildings and such. By the second semester, though, I'd lost interest and had also ruled out the chemical and electrical sciences—the mathematics described an unseen molecular world, and I needed a visual connection. That left mechanical engineering and Newton's Laws and all those bouncing, spinning, and falling objects.

I thought about Steven and Lisa for a long time. They'd been seeing other people, and their relationship had grown muddled, having started so young. Still, they'd been in love—true-blue, vibrant, and obvious when they were together. Did her decision to put Steven in the rearview mirror free her or did it make things harder? Maybe both? She was barely 18.

Over the next year, Steven worked on his recovery with the appropriate sense of desperation. He learned to feed himself with his left hand and to pull his wheelchair around the house with his left leg. Two major victories.

His right side told a darker tale. The asymmetry in his smile remained. The arm inched along as if filled with lead. Given ample time, he could flex his fingers, and every once in a great while he'd reach up and shake my hand, the pressure from his fingertips so light it almost tickled. Doctors had said that his right hand had saved his life, taking the brunt of the blow away from his head, and it seemed to remind him of that sacrifice daily. The wrist with the plates and screws inside became his greatest source of pain.

Despite the frustrations in his recovery, his right leg showed promise. Like everything else, it moved as if underwater, but he'd gained some strength and hadn't plateaued in his therapy exercises.

So, he continued to work, clinging to every scrap of progress, and we continued to wait and hope as there was nothing anyone outside of his family could do. There were plenty of offers, including the local Lion's Club chapter, who made it known—anytime, anywhere, anything—but I don't think the Botts ever asked. The only help that I knew of was from my father, who worked for the Social Security Administration and personally managed Steven's disability claim.

Alongside Steven's tragedy, my father's mysterious decline continued. We played less racquetball. He slept more—two or three-hour naps during the day, after ten hours of sleep at night. No answers from doctors. No treatment plans. Some said it was all in his head.

In my head, frustration festered, and helplessness started paying rent. School kept me busy most of the time, but when it didn't, I spent extra time at the gym. Quiet moments were always the hardest.

At WPI, mechanical engineering students could specialize in the biomedical field. Forget robots, cars, and airplanes, the human body was the slickest machine on the planet. The heart pumped, kidneys filtered, and lungs transferred gases. With this perspective, the body could be repaired like a machine, and science fiction became medical prescriptions: titanium joints for arthritic knees, insulin pumps for the diabetic, hearing aids for the deaf, and more.

Was there a cure for my father or a new wrist for Steven? I'm sure I imagined it, and in that imagining, I had found some sort of path among the weeds.

-gump-

I remember being blown away by *Forrest Gump*. The film was long but so good I didn't notice my butt losing all feeling until I stood up at the end. I saw it twice, enjoying it more the second time. *You have to see this!* I said to my father, the excitement pouring out of me like a fire hose. *You'll love it! We'll see it together!* He said yes. He was in his late 40s, the chronic fatigue worsening. Racquetball games were no longer possible.

We went to the Elm Draught House in the center of Millbury, a beloved old theatre built in the late 30's and then converted into a 300-seat, single-screen cinema pub in 1983. There were bench tables in front of the seats for popcorn, soda, pizza, and beer, and you could ask for an extra seat cushion at the ticket booth if you needed it. Concessions cost a fraction of those at the larger multiplex theaters. Many high schoolers experienced their first kiss in the back row of that theater.

Anyway, Forest grew up and made the Alabama Football team, and my father and I laughed as he ran out of the stadium with the ball. Then, Jenny was struggling, but Forest kept popping up to help. We were sucked in. Then, Lieutenant Dan was warning Forrest and Bubba about their socks on the beach, and I was a little uneasy because it was Vietnam. I relaxed, though, when then they started talking about Bubba Gump Shrimp.

How could I forget about the ambush in the jungle? My father jumped up in his seat like he was going to bolt from the theater, his movements startling me and others around us. He wasn't at the Elm anymore. He was back in Phu Bai at the Eighth Radio Research Field Station, and I was a complete idiot.

My heart dropped into my stomach. For me, the ambush had been scary but thrilling, an escapist caffeine jolt to a quiet suburban life. For my father, it was a trip back to hell, dropping to the ground, diving for cover.

I squirmed in my chair as Forrest hauled the guys out of the jungle. *Had my father seen any of his friends die in front of him? What in God's name was I thinking?*

"That's the only fight scene," I whispered when the action settled. I was sweating and sick to my stomach. I told him I had to run to the bathroom. Thankfully, Forrest had been "shot in the butt-tocks" to alleviate the tension. I stood in the back of the theater near the concessions for some time.

At the end of the movie, the blue feather floating away peacefully, I glanced over at my father. He seemed lost in the magic of the story. I didn't say anything.

"Hmm…" he said softly to himself. "Jenny had her own Vietnam."

-wedding-

Three years after Steven's accident, the miraculous occurred. Joe Bott married.

John, Steven, and Joe were the ladies' men of the neighborhood, with Joe being the smoothest of operators. "Joe should be a politician," I'd say to John before I understood what politicians did. Joe was handsome but not like Steven and his slim, angular features. Joe's visage was softer, a smile always waiting behind friendly, brown eyes. His persona matched his appearance: approachable to women and comfortable in approaching them, confidence fitting like a tailored suit. Without really knowing why, I assumed he'd live the life of a Don Juan or maybe a Sammy Malone from *Cheers*.

The wedding took place at the St. Brigid Church. Father Markey conducted the ceremony. Fred served as the best man. John and George were groomsmen. Steven was not in his wheelchair.

I remember the procession, the low murmuring, the energy in the building shifting, palpable. I raised up on my toes to see over the crowd, the sun shining through the stained-glass windows, bathing everyone in ruby, gold, and emerald tints.

There in the back entryway, Steven stood tall and impressive in a tuxedo. I'd forgotten about his height as it had been three years since our last basketball game together. If I didn't know better, it looked like he was about to walk down the aisle.

The day the doctors unscrewed the halo from his skull and removed the breathing tubes, Steven had begun a private war on his limitations with each battle providing varying degrees of success. Some improvements, like feeding himself or enunciating, were noticeable. He handled soda bottles, utensils, and slices of pizza with steady, if not, perfect control. He enjoyed playing cards, and, although someone had to help him with shuffling, he could deal the cards one at a time. Most tasks were completed with his left hand, but the right helped when it could, such as holding a paper plate while he ate.

Other improvements, like core strength, were not so apparent even though the water therapy, stabilization exercises, and abdominal work had built a foundation, one muscle fiber at a time. Reclining became sitting, and sitting eventually transitioned to standing. I'd seen him get in and out of his Pathfinder, a cumbersome process: standing up from the wheelchair, inching around until his back faced the passenger's seat, and ducking his head and using his left hand to hold onto the doorframe as he dropped slowly into the seat and then twisting his body while lifting his legs into the car. He had to use both hands to pick up his right leg.

Despite this progress, I'd never seen him walk and didn't think it was possible. He came to several of my basketball games at WPI during my senior year but always in the wheelchair. Supposedly he had tried using a walker and learned quickly that his right arm wasn't strong enough to support him.

Had something changed while I was at school?

Steven looked up at the 30 yards of aisle before him and exhaled. A small first step.

What if he falls?

From the get go, the right leg clearly lacked the coordination to smoothly transition the weight. He swung it forward in an exaggerated torquing of the hips, the movement coming from his lower back. Once his foot landed, he willed every muscle in his leg to contract—*Hold! Hold!* The hip extensors and glutes strained to pull and propel his body forward, causing him to sway and rock while he hurried his left leg in front to retake the load. Quick-slow. Quick-slow. A limp like the staggered gait of a poorly matched prosthetic limb. More an exercise of will rather than ability.

For the next shaky, wobbly three minutes, no one breathed. When he finally joined his brothers, his cheeks were flush, and beads of sweat glistened along his forehead as the whispering spread throughout the church. I wanted to clap, but out of respect for Joe and Karen (his bride) and fear of Father Markey, I refrained. As if the day hadn't already been special!

Later at the Spencer Country Inn, the DJ introduced the wedding party one at a time to the guests. When it came time for Steven, standing tall and grinning in the doorway, our enthusiasm, bottled up like a shaken soda can, erupted into a thunderous ovation.

-birthday-

On my way to a birthday party for Steven (I forget which birthday), I stopped at my father's new home to pick up a gift.

My father had left the duplex next to Aunt Jan and Cousin Rob in the early 90's and bought a small ranch in Millbury right up the street from the wire mill his father worked at for 30 years.

The home security action figure militia had some fresh recruits. Wolverine stood between a tank and a military helicopter on a shelf in the kitchen with great sight lines to several windows and the front door. The green plastic Army men and a few passenger planes covered living room windows facing the front yard. I don't remember seeing Han Solo anywhere; maybe he was in the bedroom. Most of the furniture was from his apartment. The house was cluttered but still had that familiar smell that reminded me of his parents' home.

My father was getting to be quite the collector of weapons: bayonets from Korea, Vietnam, and World War II; hatchets; nunchucks; compound bows; switchblades; and bowie knives. Look at this, he said one day, his normally sleepy eyes vibrant as he drew a samurai sword from a sheath on his belt.

The birthday party was just the eight of us: brothers, parents, Steven, and me. Maybe it was only seven, I can't remember if John was there or still in the military. Steven opened a few cards and a couple of presents. There was cake, but I don't remember the flavor. We were definitely in the kitchen. Steven sat in the chair closest to the refrigerator, his brothers and father sitting next to him. I can still see Mrs. Bott at the counter putting candles in the cake.

I handed Steven his gift, a cane, unwrapped, which everyone considered without a word. Steven owned a classy, ebony-stained wooden cane with a golden polished handle and rubber-tipped base. What I had presented to him was plain, worn, unadorned, chipped, and scratched in several places. Steven studied it, uncertain.

"Thahnkss," he said finally, and I almost laughed.

"Hold the shaft and pull the handle," I said.

After several tries, George held the cane for Steven as he pulled the top, freeing the metal core from its sheath. Yet another thing I took for granted and assumed he could do on his own, but no one seemed to notice. Steven frowned at the long metal object. There was no edge to the blade, just a point at the end, which was also quite dull.

"It's a sword," George said.

"Oh, no!" Fred said.

Steven's eyes widened, and he waved the sword around, jabbing and poking at the air above the table, chest swelling. The brothers and Mr.

Bott egged him on. Wow, look at that, Steven! Nobody will mess with you! Mrs. Bott came over from the counter, watching her son, who was giddy by that point.

She turned to me, hands on hips. "Christopher Lawrence Richards! Are you CRAZY?"

-maximum-

I played in the Millbury Summer League for nine years after that first game with Steven in high school. During one of my last summers in the area, the stars aligned, and we managed to have all the Bott and Richards brothers close by. John and I shared an apartment in Worcester; I had returned to WPI to get a master's degree; and John worked at a local bank as an account manager. George lived in the next town over with his girlfriend. Joe and Fred were both in Millbury, raising their children. Bry was home from college for the summer. For the first and only time, we all laced up our sneakers united.

Bubba managed the league and, seeing that we had signed up together, added Steven's name to the roster as an honorary captain. Steven had been walking non-stop since Joe's wedding—step by shaky step, quick-slow. Sometimes, he used a cane (the nice one, not the sword) when he was tired or during the winter when the snow and ice blanketed the ground. But he always walked, never used the wheelchair.

With ambulation came independence. Daily activities—going to the bathroom, taking a shower, getting something out of the fridge—had become somewhat routine again. Everything occurred in slow motion, but he could navigate his home and restaurants and have a meal with friends, and yes, he could attend summer league games. Mr. Bott settled into the role of chauffeur, driving Steven everywhere and always using the Pathfinder.

Steven would sit with us on the bench, cheering, taunting the opposing players, and flirting with women in the audience. Public relations, he liked to call it. With him present, we won every game, making it to the championship, which we also won thanks to a clutch 3-pointer by Fred.

After the game, we laughed, high-fived, and relived Fred's shot, which grew further from the basket with every retelling. Family and friends indulged and teased us. No trophy, no prize, just a set of memories to recall decades later in front of our bored children.

I remember someone suggesting we head to the pub up the street for more celebrating; nobody wanted the night to end. As we grabbed our things and started back to our cars, I saw Steven on the far side of the parking lot, slowly limping alongside his father, and everything turned grey and empty inside. He hadn't stuck around, and he wouldn't be joining the team for drinks. We'd be in our cars and long gone before he ever reached his. Time flowed differently for him.

Insurance companies will sometimes use the term, *maximum medical improvement,* to describe a client's final plateau in their recovery. It allows them to calculate claims and settlements, earning potential, lost wages, pain and suffering, limitations, and impact on lifespan. Although I don't know what the doctors were telling Steven, I'm sure they knew if his body had done everything it could to heal and when it was time to adjust treatment strategies from rehabilitation to management. Brains do most of the healing in the first year after an injury. Steven was in his fourth.

I didn't end up going to the pub that night.

-along for the ride-

At some point, I'm not sure when, my father and I talked about the golf ball-sized lumps. There were at least ten along his forearms. I might have thought they were muscles when I was younger.

He said the lumps were from weed killer, which made no sense. Vietnam had jungles, he said, great for hiding the enemy so they sprayed the weed killer on the nearby plants, wiping them all out. No more hiding spots.

"Why did you get the lumps?"

"Sometimes, we would get sprayed, too."

"What was in them?"

"Probably weed killer."

I remember him speaking softly, calmly while thoughts buzzed in my head like wasps. The more he talked, the more I understood that I'd never understand. Vietnam had raged 20 years ago, thousands of miles away, and yet somehow lingered.

"We had 11 good years," my mother had said. "Then five tough ones."

Sometimes, I wondered about the five tough ones, but the more I learned, the more I wondered about the 11 good. How? What else from Vietnam had come back with him? With everyone?

"You never suspected any trouble between your parents?" A psychologist would ask me years later.

"No, nothing." I said.

She frowned. "You seem like a pretty observant guy," she said.

I asked her if she knew anything about poker players.

-crossroads-

It ended in my boxy silver Toyota Camry, soon after a graduation party. She was 23. I was 25. She had an entry level job at an insurance company, a Dodge Omni, and an eagerness for the next chapter: husband and children. I loved her, and I loved kids, but they weren't on my radar. I needed answers and reassurances. A family assumed much: love, commitment, marriage, health. So fragile. So many guesses. A family assumed faith.

We faced forward and stared ahead or at the floor, anywhere but at each other. It had been 18 months, an easy glide. Work and school and then weekends together. For her, that was enough time to know that we were good.

"This hurts so bad," she said, a hitch in her breath that cut through me like a scalpel. I was frozen; eight divorces among my parents, aunts, and uncles, and the kids got kicked in the teeth with the breakups. Childhood up and left in a rush. She didn't know this—two parents, two siblings, two cats, a house, and a dog.

I remember hitting my dogs when I was in junior high school, and it still makes me wince. Ginger and Strider, a pair of small Dachshund mutts would howl in the middle of the night because we confined them to a small porch. I guess they wanted out, or maybe they'd heard another dog or something scurrying around the yard. I'd wake, blood boiling, and fling open the porch door. They'd already be cowering. *Shut up!* and a solid whack across their backs. *It's two am!* Whack! An hour later they'd be at it again. Did I ever slap them on the head? The nose? I don't want to remember. We eventually let them in the house at night, and the howling stopped. They never greeted me without fear after that. Their tails wagged, happy to see me, but always curled between their legs, reminding me of those nights for the rest of their lives. Ginger lived to be 16.

Soon after, or maybe right before, the breakup in the boxy Camry, I was accepted to two Ph.D. programs in biomedical engineering, one in Philly, the other in New York. Both degrees came fully funded, and the one in New York was supported by a foundation that worked with Olympic athletes. I'd won the academic research sweepstakes. Twice.

The interview in Philly was an unmitigated disaster. The mail I'd received from them described the school's biomechanics program and listed different areas of research, including the spine. When Professor So-and-So interviewed me, she mentioned animal studies. Then, she mentioned neck injuries, and the conversation headed into swamp water. Animal experiments? Would I be breaking necks? Severing spines to measure God-knows-what? Animal rights groups were not happy with their department she told me, so I could only imagine the research. I remember primates being mentioned at one point, and my insides started trembling. I started sweating. I'd struggled with using rats for my master's thesis research, so much that someone else had to euthanize them. They were big, white, ugly rodents, but their bodies were still warm. soft, and forever asleep because of me. At some point during the interview, I froze during one of the professor's questions, and she could tell something had gone terribly wrong. She picked up a brochure, and the conversation became an infomercial about the school to kill the time.

The visit to New York went much better. Knee research and cartilage studies. No animal experiments. They used cadavers instead, and dead people weren't a problem. I don't know how I knew this since I'd never worked with cadavers, but dead was dead, I thought, just recycling parts. The interviews went swimmingly, and the students were impressive. For the last meeting of the day, the Department Chair sat with me in his office, smirking at my Worcester accent. I asked him about animal studies, and he said there wouldn't be any for my project, "but in this field, you're going to have to work with animals at some point. It's unavoidable in medical research."

I'd chosen the wrong path.

When I returned from New York, my father said he could help me sell the boxy Camry since I wouldn't need it in the city. I told him I wasn't going; I couldn't do it. After a pause, he said something strange, seemingly out of left field.

"That's okay, we Richards are born to underachieve."

When I told Steven about the schools, he said something different.

"Ih'll pahyyhou toghotho Phillhy."

If only I could have done that for him.

The dust had settled on Dad and Steven, and I still had to fix them, somehow. Why the hyper-responsibility? The guilt? I was terrified of what they were going through, and that fear kept me holding on tight. I couldn't let go.

So, I ran.

-reincarnated-

My father once told me that karma can also be from past lives, which meant we were talking about reincarnation. We had lived in a community of Christians of all denominations—Lutherans, Episcopalians, Baptists, Catholics—and we were all going to Heaven if we played nice. Not dear old Dad; he was coming back as a frog.

"No, a woman," he said.

A woman?

"We switch genders every time we die."

He was afraid of being a woman; I think this scared him more than hell scared most Christians.

"So many hormones," he said. "It's much easier being a man."

He'd given up on the Catholics and the Christians after Vietnam. Too many castles in that kingdom he said, and the whole *The Lord works in mysterious ways* fell far short of explaining why bad things happened to good people. In his experience, very bad things.

My father followed the teachings of Eckankar, a spiritual path of light and love led by Sri Harold Klempf, the living Mahanta at the time. Eckists chanted Hu together and told stories about individual growth in this life and past lives that some remembered. Each life was a journey from which to learn.

What about a turtle? I'd tease. Or maybe a fox? I liked foxes.

He'd chuckle. Not usually, he'd say, unless we went backwards in our karma.

In 2004, at his 40-year high school class reunion, he apparently opened with this belief while catching up with old acquaintances who were likely expecting to share stories of kids and grandkids. My mother and Paul, my stepfather, were there too, sitting at the same table. Paul, who's a little hard of hearing and unfamiliar with reincarnation turned to my mother.

"Did Larry just say he's going to become a woman?"

-garden state-

I pulled into the Wayne Valley High School parking lot a few minutes to game time. Which was a problem. Long gone were the days of stepping out of the car and right onto the court. At 34—or 70 in basketball years—something in me would snap like an old elastic band if I didn't warm up first.

I was late because of the storms. Not that day, sunny and pleasant, picture perfect. No, the storms had come through earlier in the week. Nothing biblical, just cool steady rains, moseying through the region and providing a brief respite from the sweltering New Jersey summer.

But then there was the Passaic River, drinking up the runoff. It rose quietly and crept into parking lots and roadways and spread into backyards. Route 46 passed by the river twice, the eastbound side submerging at one crossing and the westbound side at the other, three miles away. The Willowbrook Mall parking lot became a duck pond, preventing access to my favorite Chipotle next to the AMC Theater. Wayne, Little Falls, and Fairfield shouldered the brunt of the mess. I worked in Fairfield.

A new job had brought me to New Jersey—"the sixth borough of New York," a friend told me—and she was right—a suburban sprawl on steroids. Seven-mile commutes could take an hour, with traffic running every which way, not just in and out of the cities. Industrial parks, car dealerships, strip malls, and commuter rails were thrown together amid lovely towns with picturesque commons. Some of it was old, some new, but there was always construction tearing things up and leaving nails in the road for my car's tires.

I didn't know a soul when I relocated but became infamous among co-workers after being introduced to them as a Red Sox fan. New employee announcements in 2004 coincided with the Sox winning their first World Series in 86 years, and my company greeted me with a hearty round of boos.

Friends back home laughed when I told them I was sold on the Garden State, but I laughed when a winter storm dumped a foot of snow on them and we got rain. Within months, I purchased a one-bedroom condo in Verona—a quaint little town with the same population as Millbury squeezed into half the land. The second-floor unit had refinished hardwoods, a closet-size bathroom tiled in a grotesque pink and black, and a set of windows along the kitchen and dining area that captured the sunset year-round.

Somehow, it all worked. Among the diners, Yankees fans, and 24/7 traffic, I'd found space and possibility; my breaths lengthened, taking their time as if my lungs had moved into to a bigger ribcage. I woke each day with a gentle nudge from the morning light. After 30 years in Massachusetts, the change provided a separation from that old, heavy stuff that I'd been too close to see anymore. With the new scenery, I'd found different patterns, different perspectives.

"You moved to New Jersey by yourself?" several friends asked, incredulous.

"Yeah!" I exclaimed.

It was a perfect fit.

Almost.

"Why?" I asked my co-workers the first time the roads flooded, trapping us at the office until eight o'clock.

"It always happens when it rains like this," they said as if describing African monsoons.

"Can't they just build up the areas and roads near the river?" We had the same weather in Massachusetts, after all.

"Don't know. Just stay away from the highways."

Wonderful.

On flood days, most people worked early or late to avoid rush hour, while others telecommuted. Some took the day off. Everything shifted and waited, and there seemed to be a choice: adapt or wring your hands and bitch to others and show up late for everything. Whatever your decision, the water left in its own time.

I was fortunate. My commute to work ran south of the flooding so I could avoid the highways and cut through neighborhoods. But to get to Wayne Valley High School, I needed to cross Route 46, and the water caught me by surprise. All I could think as I hurried into the school was that it should've been gone by now.

-turning point-

The last day of fighting of the Tet Offensive occurred on February 24, 1968, with an obvious winner to the conflict. The communists suffered ten times more casualties than the United States and South Vietnamese (National Archives, 2020); they failed to control any of the areas they'd captured in the opening days; and the hope that the conflict would ignite a popular uprising against South Vietnam's government never materialized. In addition, the Viet Cong, which had come out into the open for the first time in the war, were all but wiped out (History.com, October 29, 2009).

What should have been a staggering loss for the North, probably won them the war. The intense escalation in fighting had crushed the hopes and morale of the United States that victory was near; the VC and North Vietnamese seemed ready to drag this out to the very last soldier.

On February 27, CBS Evening News anchor Walter Cronkite, having just returned from Vietnam, gave a highly critical editorial during his television special "Who, What, When, Where, Why?" urging America to leave Vietnam.

> ...not as victors, but as an honorable people who lived up to their pledge to defend democracy and did the best they could.
>
> It seems now more certain than ever that the bloody experience of Vietnam is to end in a stalemate. To say that we are closer to victory today is to believe, in the face of the evidence, the optimists who have been wrong in the past.

After hearing about Cronkite's report, President Johnson allegedly responded, "If I've lost Cronkite, I've lost Middle America." (Merry, 2012).

By the end of March, President Johnson was addressing the nation: the United States would begin de-escalation in Vietnam, halt the bombing of North Vietnam. and seek a peace agreement to end the conflict. In the same speech, he also announced that he would not seek reelection to the presidency, citing what he perceived to be his responsibility in creating the national division over the war.

Meanwhile, in Saigon, Larry was given malaria medicine to take regularly as long as he was in the country. No questions asked. He was concerned as he seemed to be sensitive to medications, having been hospitalized during basic training after a reaction to a flu shot. He never contracted malaria, but as soon as he started the medicine, he made frequent and sudden trips to the bathroom for the next five years.

Not long after Johnson's address, Larry was finishing up in the bath house when a thunderstorm came through. He was leaning in the doorway, watching the jagged streaks of light cut across a dark sky, when a bolt struck the transformer across the street. The current traversed the power cable connected to the building and through the metal door frame where one of his hands rested. He woke an hour later, shaking and twitching, a medic kneeling over him.

"Lucky," the medic said. "Your rubber-soled boots saved your life."

Larry shook for the next two weeks, twitching while typing out his reports, his CO refusing to send him to the hospital.

Over time the twitches clustered into a singular convulsion that occurred several times throughout the day. I've seen it many times. His eyes lose their focus, his right shoulder drops a little, and then a strong spasm fires through the right half of his torso. It looks like someone is hitting him with a taser. And then it's gone.

He would need anti-seizure medication for the rest of his life.

-twilight-

On Monday nights, my knees hated me.

Basketball is a young man's game, fool, they'd say after the pick-up games at the Caldwell Community Center. *All this front-back, left-right, up-down. Do you know what stop-and-go traffic does to a car? What if you tear something?* Then, they'd hum their sad melody of cartilage long gone. A bag of frozen peas and a week of rest was the only way to shut them up.

Next to the community center stood a small, unassuming brick building facing west. I noticed it one evening walking to my car, looking up and around at a magnificent twilight sky. Crimson and gold stretched across the horizon, chased off by the darkening east, and the buildings and roads glowed in the dying light. I was thinking about how the suburban congestion and its endless concrete and machinery seemed insignificant beneath the sky. How effortlessly the setting sun washed away all of our hard-earned ugliness every night.

The small building had a small sign next to the front door with a large Chinese symbol on it. Below the symbol, was an English translation: Tai Chi. I knew something about Tai Chi, a moving meditation, healthy for all ages. A couple of months before relocating to New Jersey, I'd come across a studio not far from my apartment outside of Boston. A flyer in a laundromat claimed Tai Chi calmed the mind and reduced stress. Soon after I found the flyer, my employer laid off everyone in the company and closed its doors for three months. I decided a stress reduction practice would be a good idea.

The instructor, Peter, a stocky man with a warm smile and soft voice, had a curious background: a Ph.D. in Biology and a job researching the benefits of Eastern medical practices. He worked with several Boston hospitals and published articles in scientific journals on his findings. Conditions such as COPD, depression, and chronic heart failure all showed improvements with patients who practiced Tai Chi.

Peter's teaching style focused on health and "good vibrations," and he incorporated visualization meditations in addition to the basic Tai Chi exercises. We would close our eyes and "smile into our spleens" or "inhale energetic white light into our lungs." Pretty flaky stuff, but Peter always grounded his teachings with science.

"The mind and body have a strong connection," he'd say. If we filled our heads with negative thoughts all day, the body would follow suit, pumping out stress hormones. But if we focused on the positive, and on healing, our bodies would respond in kind. All of this could be measured in a lab.

When it came to Tai Chi, Peter taught us to "pour the weight" from one foot to the next and to "hold the ball," rounding our arms, while we moved through the different positions. He told us that we were learning a powerful martial art style of self-defense. We laughed. With our sloth-like movements, the only attacks we could have defended against were from each other.

He also kept the atmosphere light and fun. In addition to feel-good meditations, he frequently inserted jokes into his teachings.

"Chris, what did the Buddhist say to the pizza guy?"

"I don't know."

"Make me one with everything."

I groaned. "Of course, he did."

Three months after joining his class, I was offered the job in New Jersey. When I told Peter the news, he closed his eyes and put a hand to his heart as if losing someone dear to him. He then gave me a hug and wished me well.

When I signed up at the Tai Chi studio in New Jersey, I hoped to find more of the same. More kindness. More health. More good vibrations. I did not.

-intelligence-

Four months into his tour, Larry was sent to work as a support clerk and courier at the Eighth Radio Research Field Station in Phu Bai in the northern region of South Vietnam, 38 miles from the border.

The primary function of the base was intelligence, intercepting communications to learn about enemy movement and strategy. Interpreters, fluent in Vietnamese, Chinese, and Russian listened on different bandwidths 24/7 in 12-hour shifts. Where were the VC? The North Vietnamese? What were their plans? When the specialist got a hit, they never used radio communications to contact the troops in the field, afraid that the enemy would intercept their messages. Instead, clerks typed out reports and locked them in a briefcase, handcuffed the case to their wrist, grabbed a pistol, and headed off to the field to deliver the intel.

"The pistol was just for show," my father said. "All scarred up inside the barrel. I couldn't hit anything farther than five yards away."

Larry mostly traveled by helicopter or by C-130. When he flew by copter, he wished that he had a steel plate to sit on like the window gunners as the bullets rained up from the jungles into the sky during the flights. When he flew by C-130, the plane was often filled with South Vietnamese farmers and their chickens and pigs. Most of the farmers' stomachs couldn't handle the flying. Between the livestock, manure, and vomit, the odor was indescribable.

He once told me about a big score: a large VC battalion with impressive weaponry setting up to attack their base. The intel was spot on, and the U.S. soldiers were able to wipe out the VC with minimal losses.

"We sent a truckload of beer to the soldiers to congratulate them on the victory," he said, and I could tell he still felt good about helping the guys on the front lines.

Then, he added, "To thank us for the intel, the marines sent back eight severed ears from the enemy."

-chi flow-

My new Tai Chi instructor had an impressive pedigree: student of the legendary Cheng Man-ch'ing. Cheng, who started teaching in China in the '40s, became the Tai Chi O.G. of the West when he moved to New York in the '60s. Cheng had actually created the Yang Short Form that Peter had taught me in Boston.

Cheng's pupil, my new teacher, stood tall and thin with pale blue eyes, long white hair, and a flowing beard to match. A Saint Nick on Slimfast. When still, he was unremarkable, but once he moved, he transformed into something more impressive than any of the athletes I'd seen on a basketball court. There was an ease to everything—opening a door, tying a shoelace—that made the ordinary poetic, like a wave that had escaped the ocean and disguised itself in human form.

Despite his remarkable abilities, he failed to impress with his teachings.

"Hit me," he said to me one day.

I blinked, not understanding.

"Go ahead." He turned sideways, offering his arm as a target.

I glanced at Adam, his most senior student. Adam shook his head slightly.

"I won't hit back." The teacher's eyes glowed like a four-year-old concealing something in his hands.

I made a fist and struck his arm lightly, almost playfully.

"That's not a hit," he said. "Hell, Tara can hit harder than that."

Tara, also a new student, shot him a look.

"Hit me!" Something more than playfulness behind his voice.

I struck again, solid and direct, maybe 75%, half-cringing in anticipation of the inevitable ninja strike.

He never moved, but the contact pressure between my hand and his arm grew and intensified. My knuckles cracked and popped as tiny bones and joints shifted inside. Then the force translated into my wrist, heading toward my shoulder, my hand telescoping into my arm.

I yanked my fist away.

"That's better!" he said, smiling.

"What was that?" My voice stopped just short of yelling.

He laughed.

I massaged my hand and looked around the room, hoping someone could tell me how his arm had just punched my fist. The others watched me closely. Adam frowned.

"I didn't do anything." The teacher shook his arm to show me: normal, thin, unimpressive. "Go ahead," he said and held it out for me to examine.

I touched it tentatively, noting the spot where I'd struck. No bulging biceps. No titanium exoskeleton.

"See?"

I saw nothing.

He winked, and his arm relaxed in my hands and became deadweight, like a towel soaking up water, increasing from five pounds to 20. I jumped back as if it were a snake.

He roared with laughter, his shoulders shaking.

Who the hell is this guy?

Yes, this old teacher displayed the incomprehensible, and I honestly believed he didn't have a touch of arthritis despite his age. Why would he? He was water. Water didn't suffer arthritis. The magic he'd tapped into also lit up his eyes and skin. If he dyed his beard and hair, who knows how young he would've appeared.

But.

As much as his body flowed with chi, his mind seemed as dense as a stone. One would think someone of his abilities would've been a wise old sage, like a Gandhi or at least a Mr. Miyagi. He was neither. In addition to embarrassing his students for amusement, he barked orders and snapped at his wife who assisted with classes. We also suffered through an endless supply of inane theories: Hillary Clinton's fat ankles precluded her from being a competent leader; cigarettes didn't cause lung cancer; the FDA was conspiring against tobacco companies. At first, I suspected teasing, and I searched for a half-smile, but he was dead serious.

So, it may not have been a coincidence that his mind seemed to be the one part of his body failing.

"Everyone stop," he said one day, and we all turned to face him.

"Watch." He demonstrated the move many of us were struggling with and performed it with the grace of a ballet dancer.

He pointed at me. "Ah … Ah…"

"Chris," his wife said.

His wife knew enough Tai Chi to lead her own class but seemed content to help. A small Chinese woman in her early 60s, she projected kindness and overflowed with encouragement. I had no idea how or why she put up with her husband.

"Chris," he continued. "Swing your hand up and away from your body like this. Why don't you stand over there so you can watch … Ah … Ah"

"Adam," his wife said, concern pinching her face. Adam had been studying with them for ten years.

During the next month the teacher repeated himself on several stories, and the students took note. We looked at each other. *Uh, oh*! An ageless body and a sickly mind, what would happen then? Something was definitely not right with this guy.

Yet I followed, despite the warnings going off like grenades in my head. I didn't care about his pedigree and parlor tricks. No, I wanted to know how he moved and glided. Whatever it was, it would undoubtedly help my cranky knees. I'd dealt with difficult coaches in the past and learned despite their challenges. I could handle his baggage.

That was the plan. That's why I stayed.

Well, that was one of the reasons. The other was Tara.

-rockets-

The problem with the Eighth Radio Research Field Station was the tower. One hundred feet tall with blinking red lights on top. A giant bullseye for all the world to see and all the VC to aim their rockets.

Twenty-seven attacks, my father said with certainty. Not 25 or 26 or about 30. Twenty-seven rockets, six feet long with a hundred pounds of C-4 explosive in the nose had landed within a hundred yards, as if he could recall each one distinctly. This from the same man who often misplaced his car keys or headed off to Walmart without a wallet.

"Did you have any warning?" I asked. "Was there some sort of sound before they hit?"

"There was a guy," he said. "He somehow knew ten, maybe 15 seconds before it happened. We'd be playing cards in the barracks and out of nowhere he'd yell *Incoming!* and we'd run for cover. He was always right."

"How'd he know?" I asked.

"I don't remember. I don't think he ever told us," he said. "But I was really sorry to see him go when his tour ended."

I remember my father spending most Fourth of July's sitting in his worn recliner, headphones on, listening to Christopher Cross, Harry

Chapin, or Cat Stevens. Sometimes, he'd be sipping at a glass of Southern Comfort as the fireworks rang through the night.

"At first, it was the VC launching the attacks," he said. "But we could pinpoint the location from where the rocket came and send in the choppers. Eventually, the VC forced the local villagers to fire the rockets. Do it or we kill your wife and children sort of thing. Our patrols would find some hapless farmer chained to a tree with the weapon in front of him."

One day, the sun high in the sky, a large group of soldiers were walking toward the mess when the bombs came. Everyone dropped to the ground. While lying there, the thunder everywhere, Larry looked up and saw a small stream of blood spurting like a fountain from a man's neck a few yards away. A piece of shrapnel had caught the soldier just right, nicking an artery. Larry crawled over, staying low, head down, and reached up, pinching the small opening shut. Maybe 30 seconds later, the bombing stopped. Larry yelled for help. The man survived.

"That's amazing!" I said.

"My first thought was that I've got to get to that guy," my father said. "Immediately after that I thought, they probably wouldn't shoot another rocket at the same spot."

-first words-

"So, cigarettes, huh?" Tara said to me one night in the parking lot after class.

One typically doesn't sign up for Tai Chi to meet romantic partners. And yet—mid-twenties, slender build, hazel eyes, and a dazzling smile—a bona fide head-turner. Why was she taking Tai Chi? Everyone else in the class was well north of 40, as I'd anticipated.

"An unfair rap, apparently," I said.

We'd exchanged several knowing glances and eye rolls during another one of the teacher's unprovoked digressions about cancer and cigarettes: the lack of scientific evidence, the FDA strong-arming. He must've owned a tobacco farm the way he carried on.

"Would you like a pack?" I asked. "I was just heading to the store."

"No, thanks. I'm gettin' me a pipe."

"A pipe?" I choked. "My grandfather smoked a pipe. I think you could pull it off. You'd look … grandfatherly."

"Aww, that's sweet. No one's ever said that to me before."

-the big joke-

To keep the VC ambushes to a minimum, herbicides were used around the perimeter of the base and along the nearby roads and paths. Soldiers would hack away with machetes, clearing brush, and then the guys on foot with backpacks and in jeeps would spray the chemicals on whatever was left standing. Helicopters would head out, too, make runs all over the area, and bye-bye VC hiding spots. Most of the time, spraying occurred when no soldiers were around.

My father doesn't remember the pilots' names or how many there were. He just remembers the guys in the helicopters thought it was funny to spray the soldiers at the base during their runs. On the way out to the jungle or on the way back, or both, the pilots would open her up a little and give everyone below a quick shower. Because it was funny.

Nobody else thought it was funny. Larry was pissed off. It couldn't have been a good idea, getting doused with a weapons-grade weedkiller. After the third time, he started complaining. He complained to the other guys and some of them were irritated too, but they weren't doing anything about it. He complained to the pilots, who ignored him. He complained to his captain, who did nothing. The pilots weren't his men, the captain said, go complain to the guys in charge of clearing the brush. The guys in charge of clearing the brush said don't worry about it, but go complain to the pilots' captain. The pilots' captain told Larry that the guys were just having a little fun. Nobody was getting hurt. Larry kept going up the chain of command, running into wall after wall. Eventually, Larry's CO told him to shut up and drop it.

Two months after his first exposure, Larry woke one morning to the terrible realization that he'd been right about the chemicals all along.

-poetry-

Besides Tai Chi, Tara enjoyed hiking, yoga, dancing, baseball, and anything Bob Dylan. She worked as a special education teacher in West Orange, the town just south of Verona, and she lived 20 minutes away in a studio apartment. As impressive as she was at first glance, she also possessed an infectious laugh, rapier wit, and quirky sense of humor. I never knew what she was going to say next.

Soon, we were at a Yankees-Red Sox game in the Bronx together, with Tara telling anyone within earshot that I was a Red Sox fan. Then hiking in the hills near Nyack. The trail, a little too rugged for novices, forced us to shorten the expedition and go for an early dinner at a Jersey diner. We sat in a quiet corner, peering over the twenty-page menu.

"So, why Bob Dylan?" I asked.

I knew nothing about Dylan; my parents never listened to him, nor did any of my friends. For me, the '80s and '90s meant anthem rock followed by grunge. From Def Leppard to Pearl Jam. Thanks to many of my basketball friends, I also became a big fan of Public Enemy.

"He's a poet," she said in between sips of her drink.

"I'm almost certain that I've never heard any of his songs."

She tilted her head. "I find that hard to believe."

"What kind of music does he play?"

"Mmm … maybe folk or blues. Maybe classic rock."

"Eeesh, like Zeppelin?" I groaned. "I tried to get into Zeppelin, but the whining..."

"Well, fortunately for you, Dylan doesn't whine."

"For me? Is there some Dylan in my future?"

"Maybe," a coy smile, and her eyes fell back to the menu.

-bubbling of fear-

"They found a lump."

I heard those words recently during a phone call with a friend about a family member. Words that slam on the brakes. Change the trajectory of a life. A lump in her breast. A lump in his armpit. A harbinger of tests and biopsies. Hopefully benign. If not, then treatments. Hopefully remission. Hopefully wearing the label of survivor. Hopefully.

When the first doctor examined Larry, he said that it might be "fear bubbling up from stress." Larry went to another doctor.

The next doctor had no idea what was happening, but he had seen it with a couple of other soldiers. He listened to Larry's story about the helicopter spraying and wrote everything down. Could it have been some sort of a reaction to the chemicals? Maybe. Who knew? They were in unchartered territory. Larry had no other symptoms outside of the convulsions from the lightning strike and the chronic stomach issues from the malaria medication.

All the doc could do was conduct a head-to-toe exam and document everything he observed. There were lumps—large, bulbous and fatty. And they were everywhere: both forearms and biceps, lower back, abdomen, and one small one on the face, near the eye.

Fifty-two in total.

-push hands -

In the fall of 2006, Tara and I completed the Yang Short Form of Tai Chi. No ceremony followed. No belt. Not even a high five.

"Now," the teacher said. "You just need to practice that for the next 40 years ... ah ..."

"Chris!" his wife yelled across the dojo.

"Chris," he said and grinned, a grin that grew more absent with each passing week.

Up next: Push Hands, the gateway to the martial arts of Tai Chi. In Boston, I'd once seen Peter practicing it with his senior students.

"Think about conflict with another person," Peter had said. "It doesn't have to be physical. It could be an argument. People tend to get uptight with confrontation, their energy rises, and their body stiffens."

He inhaled and held his breath, and the muscles in his neck tightened so that his chest and shoulders rose. Short breaths, flushed face, watery eyes—I knew these symptoms all too well. The blood flowed up and got stuck in the head, sometimes during work presentations or public speaking. Often with arguments.

According to Peter, that was the worst place to be during conflict, losing balance and connection to the ground where our strength resided. Beginner Tai Chi taught how to move while relaxed and *pouring* weight from one leg to the other, keeping the energy down. Beginner Push Hands taught how to maintain this base while taking on someone else's aggression, whether it be physical or psychological.

Oh man, did I want to learn more—to be able to control my emotions, to stay cool under fire. But my time in Boston had ended. As fate would have it, I was destined to learn Push Hands in New Jersey, not from Peter. I still try not to think about this point.

-nothing to lose-

Not the rocket attacks. Not the sprays of bullets during the courier runs. Not even the lightning strike. Somehow everything had just missed him. If Larry managed to finish his tour, it wouldn't be the VC or the North Vietnamese that killed him. It would be a handful of American pilots armed with made-in-the-USA herbicides, who were *just having some fun*.

Fifty-two tumors.

The next time Larry saw the helicopter heading out and flying low, getting ready for another shower, he grabbed his rifle and stood directly in its path. Aimed the weapon straight at the pilot.

-conflict-

The basic Push Hands exercise resembled a very slow and simple dance. Partners faced each other, about two feet apart. Both stepped forward with the right foot. Knees bent slightly in a mini-squat, each person raised his right arm in front of his chest, as if holding a shield. The backs of the forearms pressed against each other, serving as the only point of contact. One person then shifted his weight forward to his right foot while the other mirrored this motion, shifting backwards. They moved as far as was comfortable and then reversed direction, with the forearm contact allowing them to feel each other's intention. Movement came through the legs and hips. The feet remained stationary.

That was it, Push Hands 101. We were supposed to learn two things: stay grounded and stay connected to your partner's intention.

The instructor's wife paired up with me for my first go at it, and within a short while she had me moving with her in a slow gentle rhythm. "Feel the change in pressure between our forearms," she said. "Less pressure, advance. More pressure, retreat."

We practiced a few more minutes before the teacher made his way over to us. He watched us for a moment and then stepped in front of his wife without a word.

Here we go.

He dropped into the proper stance with ease as I worked my way back into position, trying to remember my basics. *Relax the arm. Feel your feet. Eyes*—and he was off, moving quickly. I reacted in a herky-jerk motion, and he stopped.

"No," he said, unimpressed, and returned to the neutral position.

We started again, but this time he slowed his pace for me, and I did a fair job of staying with him.

"Good, Chris!" his wife said.

After a couple of back-and-forths, the teacher's eyes rolled up and began fluttering. For a moment I feared he was having a fit. Then, his pace quickened, and I screwed up, pushing hard into his forearm and then pulling away while he shifted backwards. But he didn't stop. Instead, his motions became wilder, twisting and turning, and then crouching. I was lost. At one point, he pushed me so far back that my front foot lifted off the ground.

To hell with this!

I turned my feet, tightened the muscles throughout my shoulders, arms, legs and got low as if playing basketball.

His long white hair fell forward, covering his face, and he exhaled loudly through his nostrils. We moved at a frantic pace, twisting, dipping, and lunging. I followed though every movement was wrong. I wasn't staying relaxed. I wasn't concentrating on my feet's connection to the ground, on everything that I was supposed to be learning. It might've appeared that we flowed together, but I knew better. So did the teacher. So did his wife.

When the fit or whatever was happening stopped, his eyes opened, peeking between strands of snow-white hair, and he smiled dreamily. Every student in the dojo was watching us.

"That's how you do Push Hands," he said.

I frowned and glanced at his wife, who was staring daggers at her husband.

-buts-

The pilot saw Larry below with the rifle and veered away before letting out any of the herbicide. He turned around, landed his helicopter, and stormed off.

"I wouldn't have shot him," my father confessed. "Just a warning across his bow."

Later that day, the unit captain called Larry into his office.

"Did you aim your rifle at one of our pilots?"

"Yes," Larry said, "But…"

The captain didn't care about Larry's buts. Larry could tell his buts to the Article 15 Committee.

-ache-

The teacher didn't say anything else about the Push Hands demonstration. I was hoping for a nugget of wisdom or cryptic foreshadowing how it would all make sense to me someday. Nope. Nada.

He simply moved on without a word to help another student. I asked his wife about it, and she just shook her head.

"Never mind that," she said.

The next day, a dull ache pulsed on the sides of both legs. I rubbed the hips bones, trying to calm them.

Two weeks later, I was still rubbing.

I showed the teacher. "I think it's from the Push Hands exercises. It feels like something is straining or pulling."

"Your legs are just too tight from basketball," he said. "It will go away as your muscles loosen up."

Tara wasn't feeling it. Adam had never felt it. Maybe it was the years of basketball. As much as the teacher irritated the hell out of me, he hadn't become a walking wave without understanding what worked. Experience had taught me that I should trust him. Coach Dunham's wind sprints had set fire to my lungs during basketball practice, but I got faster. My arms screamed bloody murder after weight training sessions, but I grew stronger. Just another example of *no pain-no gain* training. The discomfort would subside. The body would adjust. The muscles would loosen up.

A few weeks later, I attended my last class.

"Are you okay?" Tara asked me on our way out to the parking lot afterward.

I was rubbing both hip bones; they were throbbing.

"Yeah, just a little sore," I said.

It was the week before Christmas; the evening presented an inky sky and bitter wind. We stopped in front of her car, and her mischievous smirk was taking shape, loading up something clever that would, no-doubt leave me chuckling. But I didn't want to dawdle or talk about the class. I wanted to get home and ice my legs.

Her playfulness receded like the surf when our eyes met.

I gave her a quick kiss. "I'll call you tomorrow."

"Bye," she said, and the concern in her voice followed me as I hurried off to my car.

-article 15-

In the United States Armed Forces, non-judicial punishment is a form of military justice authorized by United States Code Section 815 Article 15, permitting commanders to administratively discipline troops without a court-martial. Punishment can be a combination of any of the following: reprimand, reduction in rank, correctional custody, loss of pay, and extra duty. The receipt of non-judicial punishment does not constitute a criminal conviction but is often placed in the service record of the individual. Proceedings are known by different terms among the branches; the Navy and Coast Guard call it "Captain's Mast" or "Admiral's Mast," depending on the rank of the commanding officer. In the Marine Corps, it is called being "NJP'd" or "being sent to Office Hours." In the Army and the Air Force, it's simply referred to as Article 15.

Larry was livid about the hearing, certain that they were going to demote him and dock his pay at the very least. Insult on top of injury. Even with the lumps, the evidence right in front of them, there was nobody on the base who was listening.

So, he called his wife.

-more time-

In February of 2007, a month or so after quitting the Push Hands class, I met up with a doctor: heavy, balding with a ring of bushy black hair, thick steaks for hands, sausages for fingers.

"On a scale of one to ten, what would you rate the pain?" he asked.

"Rate?"

"Yeah, you know, one's a scratch, ten's a gunshot wound." He shrugged as if to say sometimes these things happen.

"Well, it's definitely not a ten," I said. "And it changes. Sometimes, there's nothing and sometimes it's maybe … a seven. It aches nonstop."

He nodded and pressed around the hip bones. "Both hips?"

"Yes."

He said that my x-rays looked okay and then pulled a small pad out of his coat pocket and started writing. "Probably a sprain. I'll give you some medicine that should help. If it's still bothering you in a couple of weeks, we'll shoot you up."

"Huh?"

"We'll give you a cortisone shot." He tore off the script and handed it to me.

I didn't want a cortisone shot. Didn't they just mask pain?

The prescribed anti-inflammatory worked a little better than the non-prescription anti-inflammatory I'd been taking. As soon as the pills wore off, though, the ache returned, sometimes worse. Nothing improved.

A month later, in lieu of a cortisone shot, I started physical therapy. The wonderful Cathy Stephenson of Focus Therapy in nearby Montclair isolated the injury to the muscles and tendons along the outside of the leg. She massaged sore spots, broke up tightness. Then, she wrapped large rubber bands around my ankles and had me walk frontwards, backwards, and sideways, exercising muscles I didn't know existed.

And the pain in the left hip went away.

The right hip remained an enigma. PT worked, and then it didn't. Two good days followed three bad. Then a wonderful morning but tough evening and horrible night. We changed exercises, changed massages, added new stretches. Sitting hurt, but standing was okay, I reported to her one week. Then walking hurt, but sitting was okay, I told her the next. Cathy took notes, asked more questions. Did I do anything differently? Not that I could remember.

The hip just needed more time, we agreed.

-congressman-

On November 4, 1984 Harold Donohue, a lifelong resident of Worcester, Massachusetts died at the age of 83. Mr. Donohue, a bachelor, was survived by his brother, James, and six nephews and was interred in St. John's Cemetery on the southwest side of the city.

Harold had graduated from St. John's High School in nearby Shrewsbury and then from Northeastern University School of Law in Boston. He served as a councilman and alderman for Worcester while practicing law for several years before heading off to join the Navy during World War II. Upon his return, Mr. Donohue was elected as a Democratic Representative for the Third District of Massachusetts to the United States Congress. He would serve in that role for the next 27 years.

In 1974, his last full year in Congress, Mr. Donohue was the second ranking Democrat on the Judiciary Committee, which considered the articles of impeachment against President Nixon for his involvement in the Watergate scandal. Before the hearings, Mr. Donohue said the scandal was "the apex of the erosion over time of the nation's moral fiber, and he promised that the inquiry would be carried out thoroughly and impartially in order to remove the blemish of the scandal on the face of the country." He offered the formal motion to impeach that the committee ultimately approved (US House of Representatives, 1974).

A few years before Watergate consumed his attention, the congressman received a call from Vincent DeFeudis, an old friend from high school. Vincent's oldest daughter, Marie, was married to a young soldier in Vietnam who was furious about an impending Article 15 hearing. When Vincent told Harold why his son-in-law was so upset, the congressman became curious.

-gifts-

In the spring of 2007, I purchased private swing dancing lessons as a birthday gift for Tara, who told her mother. Apparently, Mom was impressed.

Before moving to New Jersey, before beginner salsa and swing classes, I was a terrible date for the dance floor. I shuffled slowly in self-conscious circles during high school proms and then later at weddings.

But at 35, I found myself leading a beautiful woman through turns and slides, catching her at the waist, guiding her by the hand. The energy flowed, and the air hummed with electricity when we got it right. Even when things went comically wrong, stumbling and colliding, there was magic. One night, Tara accidentally punched me in the chest after completing a turn, the blow knocking me back and sending her to the ground, giggling helplessly.

A month later, my birthday arrived, and she handed me a small rectangular present: *Explorer's Guide 50 Hikes in New Jersey: Walks, Hikes and Backpacking Trips from the Kittatinnies to Cape May.* I ran my hand along the book's cover and flipped through the pages: Black River Trails, Jockey Hollow, Cattus Island. All unknown to me.

"Thank you. It's perfect," I said.

For our first hike, Island Beach State Park, empty sand stretched miles in front of us. Tall, grassy dunes swayed in a warm breeze on one side, the Atlantic gently lapped the shore on the other. At the southern tip, shadowy profiles of sand sharks rose and fell with the incoming surf. "They're feeding this time of day," a fisherman said as he struggled to unhook a mottled dark grey, 2-footer with a small group of us watching in awe. "Not a good time to fish," he muttered.

We arrived back at the car as the sun dipped behind the dunes, setting fire to the sky. Tara picked up the hiking book and turned to the dog-eared page. She wrote *OK* in pencil in the upper left-hand corner and circled it.

"OK?" I asked.

"My mom does this every time she finishes a crossword puzzle," she chuckled. "I have no idea why, but we should do this whenever we complete a hike."

"OK it is," I smiled at her. "Forty-nine to go."

The night of our last dance lesson, something changed in the hip. Not during the class, but afterward at her doorstep during the goodnight kiss, that moment of acknowledgement that something special was building. Terrible bone-deep waves of agony rolled in one after another.

I bent over, hands on knees. She leaned in, hand on my shoulder. The surge passed, and we slowly worked our way back to the date and dancing and to finding a club with swing music where we could try out our moves.

"But let's get your hip better first," she said. "No pain, okay?"

-message-

The day before his Article 15 hearing, Larry was approached by his CO. The message was brief, and the messenger was clearly unhappy with delivering it: the Article 15 hearing had been cancelled.

Things were about to get much worse for Larry.

-heart of the matter-

The hip didn't get better. There were dates: dinners and movies in Montclair, vineyards along the Pennsylvania border, the beach. Tara was a fierce mini-golf competitor and even fiercer when it came to Scrabble. She took me to a Nets-Celtics game, the seats five rows from the court, and a God-awful Celtics team somehow won at the buzzer on a Paul Pierce jump shot. I took her to a Knicks game in Madison Square Garden. I met her family at a July 4th cookout.

On weekends, we poached salmon or ordered sushi from Ocha in Caldwell and watched *What Would You Do?* John Quinones, that sneak, set up his hidden cameras throughout northern Jersey, but he wouldn't catch us by surprise. We'd be ready. "Sooner or later, we're going to end up on TV," I'd tell her. She made up a catchy song for the show and then others about Stonyfield Yogurt and Nissan Stanzas and confided in me that her dream was to start a company that specialized in advertising jingles. Sunday mornings we made banana-walnut pancakes or ham and cheese omelets, and she proposed that we open up a breakfast delivery service because "who wouldn't like to be served breakfast at home?"

But we were never alone during this time. Six months had passed since Push Hands, and the hip was becoming something else, an entity that tagged along and constantly interrupted with its *Helpme!helpme!helpme!makeitstop!* A little, yappy dog in my skull.

That loop took up energy and attention, drawing from my ability to be a present partner. To listen about a tough day at school or a noisy neighbor. To enjoy a quiet walk in the park at dusk. To appreciate waking up next to her. All I had for us was scattered sunlight.

Our one-year anniversary came and went. Naturally, she'd been probing about the future, hypotheticals about living arrangements, about children.

"Children?" I laughed bitterly. "I'm having a hard-enough time taking care of myself."

I could've said yes. Move in, please. Fill up the empty spaces, the blank walls and corners of the condo, and make it our home. Our kitchen. Our towering Oak in the back. Our sunsets in the dining room window. We'll figure out the hip and everything else as we go. But I flipped the sequence and put the hip first, forcing us into a holding pattern "until I got better."

And how would I get better? Nobody knew. Doc number three dropped terms like *piriformis* and *iliotibial band syndrome* and prescribed lidocaine patches, which did little. He also wanted to inject cortisone into the injury, but I refused again, concerned about risks and side effects. I journaled everything for Cathy, and we reviewed the notes and chased false leads and triggers, both of us running out of ideas. Cathy had other therapists at the studio treat me. Nothing changed.

By mid-December the underlying tension with Tara had metastasized into resentment and on a black, frozen evening, it came to a head. We sat on the bed in her studio apartment, facing each other, the dialogue growing opaquer with every word.

Helphelphelpme!

It had been a year since the Push Hands class.

Makeitstopmakeitstop!

"There's a knife in my hip, and I can't get it out," I said at one point after a long, resigned pause.

She said nothing, her gaze looking past me, her eyes growing soft and sad.

-night watch-

"I got every shit detail after that for the rest of my tour," my father told me. "Thanksgiving, Christmas—every holiday."

Some of those details included guard duty. The Eighth Radio Research Field Station was well protected by minefields, chain-link fencing (to disperse any explosives launched directly at the base), rolls of barb wire, and soldiers manning M60 machine guns behind six-foot high bunkers of sandbags.

After his disregard for chain of command, Larry drew many of the overnight assignments, which included three soldiers for three shifts: the ten-to-one, one-to-four, and four-to-seven. The one-to-four was the hardest, he said; sleep, then high alert in the middle of the night, then sleep again.

After several nights, he noticed two things: first, the other guys stationed with him also had top secret crypto security clearance; second, they were always guarding the part of the base next to the ammo dump and jet fuel.

"Not that we knew anything important as clerks and couriers, but I think the captains figured it would be better that if a sapper somehow got through our defenses, the resulting explosion from the jet fuel would eliminate the possibility of us ever being captured."

But Larry also noticed something else since his hearing had been cancelled: the helicopter pilots had stopped spraying the soldiers. A few months later, everyone at the base was ordered to stop using herbicides around anyone, including civilians. Supposedly, the directive had come from the commanding General in the region.

"I had no idea at the time, and I'll never know for sure," my father once said to me. "But pulling a gun on an American soldier may have been the most important thing I've ever done in my life."

-weekends-

Life didn't flinch. It dragged me along, and the ebullience defining those first years in New Jersey faded. I no longer enjoyed, I endured.

Cathy and I decided to gung-ho it; bull through the exercises, dig deeper into massage, stretch more. Maybe we could kickstart something. I flared for two weeks.

A co-worker referred me to a chiropractor. Some mild neck discomfort disappeared. The hip remained unaffected.

I took a month off from all activities and physical therapy. Stopped everything. Rested. I got worse.

At the office, I carried around a beige foam seat cushion, $13 at Target. Brought it to meetings, to the lab, and kept a spare in my car trunk. Sitting on any hard surface felt like a drill plunging into the back of the hip bone.

"Whaddayahave, hemmorhoids?" Frank the Ops Manager asked every week at our team meetings.

During lunch, I drove to empty parking lots, woofed down a sandwich and granola bar with some meds, and then lay down in the back seat. Afternoons were tougher than mornings, Fridays were much harder than Mondays. Each day felt like a week. Yet, another eight months passed without my conscious knowledge as the pain feasted on my attention.

This realization hit one Friday afternoon in the dead of summer while walking out to my car alongside Kevin, a co-worker. The Jersey sun shone bright and hot, and I asked him about baseball because his teenage son ate, drank, and lived for the game like I once did with basketball. Kevin had maybe 30 waking minutes that Saturday that didn't involve driving, watching, or coaching his son.

"How about you?" He asked as we reached our cars. "Any plans for the weekend?"

Weekend? Once upon a time, a weekend entailed a trip back home to visit Mom and Paul and Bry and Dad. Or a train ride into Manhattan or a day at the beach, the warm, shifting sand underfoot, sunlight dancing along the waves. Sometimes, I spent evenings in Montclair with friends,

Tikil Gomen at Mesob or Mojitos and paella at Cuban Pete's. A date with Tara. An adventure. Something.

But what had it become? A mission: 48 hours to bring the pain down before Monday morning. Two days of rest sandwiched between five days of work.

Friday nights, I lay flat on the couch, rotating ice packs, with take-out sushi or pizza delivery on the coffee table and a Blockbuster DVD playing on the TV. More anti-inflammatories with dinner so I could sleep off the ache that had built up through the week.

The days crawled by, scheduled around ice, rest, and home PT exercises. I avoided sitting as much as possible and eschewed the usual meds to give the stomach a break from the intense dosing during the week. There might have been a swim at the community center or maybe even a walk in Verona Park. Basketball was a memory. Dancing and dating had ceased. Why bother? Friends tried to get me out. Sorry, still too sore, I'd say, hoping for forgiveness and patience. Most get-togethers involved sitting, a luxury for which I had minimal currency. Instead, I'd head out for a quick trip to a mall, just to escape the emptiness of home.

Sunday night would arrive with the hip finally quiet and the stomach settled. But the isolation was growing, and I looked forward to Monday mornings with Kevin and Keith poking fun at my Red Sox. To young, sunshiny Carla, smiling her way around the labs. To Jeff and his limitless knowledge of superhero movies and comic books. To Joanne and Beth, the animal lovers who commuted over an hour and wouldn't tell us how many cats they had. I guessed seven, but the way Beth smiled suggested I wasn't even close.

Never mind the insurance claims I was racking up; work offered psychological salvation. While others sleepwalked their way through morning meetings, muttering hellos as they sipped their coffees, I hustled and limped around the building, driving my projects through inspections and testing. I reviewed marketing spreadsheets and sales projections, listened to stories from the clinical team—patients with grandchildren had survived heart attacks because of the devices we made in the labs. We sold thousands of units to hospitals, and that was important. Therefore, I was still important, I told myself, no matter what I didn't do outside of the office walls.

In September of 2008, I was promoted to a senior engineering position. After a moment of joy, it hit me. How the hell was I going to keep this up?

-hit-

When my poor grandmother mercifully passed after one of those long, horrible, draining, everybody-loses battle with dementia, my father became embroiled in a fight with the Catholic Diocese of Worcester. He paid a sizable bill to the church and learned that his mother would be buried in a shared grave without a headstone. Upgrades and bells and whistles were available, however, for additional costs.

My father and grandfather went back and forth with church officials on services and fees, never really understanding the financing. In the end, my grandmother was buried in the public section of the cemetery, nowhere near the other Catholics. My father wrote the diocese a scathing letter chastising their money-hungry ways.

Less than a year later, he was rear ended by a woman in a van while taking a left turn on Rte. 9 near Spag's Discount Store in Shrewsbury. Apparently, something had fallen in the back of her van, distracting the woman. She hit my father's car so hard, he was knocked unconscious.

Her insurance refused to pay for damages. He had to hire an attorney and take them to court.

"You were rear-ended," I said. "Isn't that a clear-cut case?"

"It should've been," he grumbled.

The day of the hearing, the defendant entered the magistrate's office in full black and white flowing regalia. Up until that moment, my father had no idea. A nun!

"That's when I knew," he said. "A Hit Nun."

"The Catholics were after you?"

"They were mad at me for the letter."

He walked away with 20 grand for damages and pain and suffering for a concussion and a herniated disk. Not much, considering the extent of the neck injury, which flared and radiated down his arm, sometimes untouched by anything short of an opioid painkiller. His hand tingled, then grew numb and weak. The nerve was getting crushed.

Google searches suggest cervical fusion surgeries have an 80% success rate of relieving arm symptoms associated with nerve impingement and a 70% success rate at helping with neck pain. Results have improved over time since surgeons learned to go through the front of the neck instead of back.

But my father could never get the surgery, no matter how bad the pain. Physically, he was an ideal candidate. Psychologically, he was a terrible one. No scalpel could ever touch his neck.

-walk in the park-

On the July 4th holiday in 2009, John Bott stood next to me in Verona Park, the nicest small-town park I've ever seen. There was a pond with fountains and a gorgeous stone bridge that attracted wedding parties like migratory birds during the warmer seasons. A boathouse served ice cream and rented paddleboats. Wooden benches sat along paved walking paths facing the pond, each looking out at a picture worth painting. Closer to the parking area, simple stone tables imprinted with checkerboard patterns waited under a shady grove of elms. Of all the crazy games we played together as kids, I can't remember if John and I ever played something as traditional as checkers or chess. If so, I'm sure we bloodied up the rules somehow.

I moved awkwardly with the crutches, a cumbersome brace across my hip—my first venture out of the condo since the surgery. The day was bright and warm, and John was patient, taking everything in while I pointed and explained. The pond had trout and kivvers and koi, or some type of carp near the small footbridge on the far side of the park. Painted turtles and a muskrat congregated at the base of the 10-foot waterfall. Sometimes, white-tailed deer, gentle and graceful roamed the canals, their dark eyes holding you for a moment before they turned a corner and disappeared into the surrounding suburbia.

I'd rarely seen John since he'd moved to Washington D.C. years ago, but he offered to help with the recovery without hesitation or even me asking. And I would've asked without hesitating. We looked nothing like the skinny, scrappy kindergarteners from decades earlier, and yet little had changed between us.

After two years of monkeying around with well-meaning, but ultimately unhelpful physicians, a local doctor had directed me to the Hospital for Special Surgeries in New York City, a top orthopedic hospital in the country. Within a few weeks they had a diagnosis and a plan.

"It looks like you have a torn labrum," the surgeon said.

He was built like a soccer player, stocky and muscular. Even in his scrubs, he looked fast. There were three others with him: the office manager, who did not look fast, a distractingly tall physician's assistant, and an intern, about whom I can't recall a single detail, but I know he was there. The labrum, the doctor explained, is a ring of cartilage along the perimeter of the hip socket.

Over two years had passed since the injury. Why was I just learning about torn labrums?

The surgeon wanted to prescribe a cortisone injection into the hip socket to see if the symptoms disappeared. If the problem was in the labrum, the injection would help for several weeks before wearing off. If something else was causing the issue, the injection would do nothing. The office manager typed every word into a computer as the physician's assistant and intern watched me closely. Their MRI showed damage to the labrum whereas images from other hospitals did not. Other doctors had wanted to blindly stick a needle into my leg and see if it helped, but this surgeon planned to use the injection to test his hypothesis before proceeding with a surgical treatment. The whole experience reeked of a higher level of care and thought.

"Okay," I said. "I'll try it."

"Good. My office will help you schedule the injection."

They got up to leave.

"Doc," I said. "I've had this injury for two years. Why hasn't anyone else found this before?"

Four smiles. "Because not too many people can do what we do," he said.

"Now that the hip is better," I said to John as we moseyed toward the waterfall. "I want to visit you in D.C."

We talked about a trip, and I began a lot of my sentences on that day with *Now that the hip is better...*

A good day.

-surgical strike-

One night in Phu Bai, Larry drew the four-to-seven shift on guard duty. Not too bad. Sleep a good part of the night, then wake up early and watch the rising sun crest the jungle canopy. But on that night no one ever woke him up. Instead, it was the morning light nudging him. How had he slept through his shift?

He found the soldier who had manned the one-to-four shift in the bunker behind the machine gun. He had been killed. A knife wound to his throat.

The third soldier, who had the ten-to-one was sound asleep, untouched.

They soon realized that no one else anywhere in the base had been hurt. Just the one casualty, supposedly one of the few soldiers awake.

Had he fallen asleep? How did someone get through the minefield and barb wire? Did the attack come from inside the base?

"The one rule about Vietnam," my father told me. "Is that everything there was trying to kill you."

To Larry's knowledge, nobody ever found out what happened that night. But that image—the soldier's throat—continued to haunt him long after in the form of night terrors. Someone was drawing close as he lay there, frozen, unable to will his arms and legs to respond. Heartrate soaring, adrenaline pumping, sweat pouring freely. A glint of steel in a shadowy hand. The dreams occurred a few times every year for decades until a psychiatrist was finally able to help him through it with something called energy tapping.

So, when it came to the Hit Nun and surgically fusing his herniated disc, my father couldn't get past a scalpel cutting his throat while he lay there asleep.

The months passed and the nerve flared and screamed for relief. His hand would tingle and lose feeling, his arm would go limp for days. He carried the arm around in a sling. He visited a chiropractor. He popped Percosets. Fear versus pain: primal impulses at an impasse. In the end, Larry chose to suffer the pain. In the end, I think the nerve gave up.

-snowfall-

Leg presses, then crab walks with an exercise band, which always reminded me of the lateral defensive movements in basketball. Cathy leaned against a treatment table, watching closely.

"Good?" she asked as I finished.

"Fine." I shrugged. "The groin's still tight, but the hip's fine."

Since the labrum surgery, the muscles in the lower abdomen held on for dear life, especially after using the exercise bike. The surgeon assured me that it happened with some patients and could be remedied with deep tissue massage.

"Stretches, and you're done." Cathy grinned. "One more week, and you're a free man."

Free, indeed. A prison sentence ending.

I stretched my stretches and said goodbye. Outside, the air had turned frigid under a darkening sky. Experts on every news station predicted the nor'easter pummeling D.C. would hit us the following evening, but so what? It had been eight months since the surgery, and life was easy again. What was a little snow?

Early the next morning, I sat at my desk in the corner of an empty cubicle farm, fingers clackityclacking along the keyboard. Fluorescent ceiling lights hummed, and hot air breathed through vents, caressing the back of my neck. The door to a nearby breakroom squeaked open.

And there, nestled among the mundane, I felt something faint. Something not right in the hip. My knee bounced under my desk.

I headed to the bathroom, and the signal retreated as I washed my face and caught my reflection in the mirror, the panic already starting to tighten the features. *Just my imagination*, I assured myself.

Coworkers trickled in and compared weather reports, enthusiastic about the approaching storm and inevitable day off from work. I sat at my desk, a light film of sweat coating my lower back.

An hour later, during a meeting I felt it (heard it) again: like faint barking off in the distance, but undeniable. My heart revved.

"Chris?" Several faces staring at me.

I apologized for blanking, and Beth repeated her question about prototypes and tests. I scanned my timeline, lost.

"It's right here on your milestones," she picked up my agenda.

The barking grew louder.

"Are we still on schedule?" Beth asked the team, her eyes on me.

It'll be okay. I just did too much at PT last night. It'll be okay.

The sensation in the hip never intensified enough to call it pain, but it hung around all day.

Later that evening, I rushed out of the office into a frosty, thick blanket of air, alive and full A few flakes dropped lazily from the sky. Criers. Heralds. It was beautiful.

I was shaking.

-interrogation-

When I was an adult, my father told me a story to help me better understand the inhumanity of war.

One day in Phu Bai, Larry was ordered to drive a Lieutenant into a nearby town. He didn't remember why specifically. The town, which was more like a village, hosted a small military base of Thai soldiers, allies of the U.S. and South Vietnam. As my father told me this, he lightly tapped the table with his fingertips.

On that day, the Thai soldiers had captured two alleged VC spies among the locals. Instead of interrogating the men in a small windowless room with a bright light shining in their faces, the Thai paraded the men into the village square in broad daylight in front of everyone. The questions were civil, even-toned. Larry didn't understand the language so he had no idea what they were asking, but he could see the Thai were growing frustrated, not getting the information they wanted.

One of the officers struck the men and started shouting at them. The villagers were appalled. Undeterred, the officer's tone grew harsher and the examination violent. Blood was spilled. The villagers started weeping. From there it descended into full-blown torture. No apologies, no effort to hide anything from the onlookers. The villagers became hysterical; there might have been family members present. Hell, the men might not have even been VC spies. Larry didn't know. Larry wanted to get the hell out of there, but the Lieutenant wanted to stay.

My father didn't reveal the details of the torture, but his fingers were tapping the table with every syllable. "After almost two hours of brutal interrogation," he said, "The Thai soldiers decapitated the prisoners and placed their heads on poles in the center of town before heading back to their base."

The image steamrolled me—jungle heat, howls of agony, blood everywhere—and I felt my body go rigid. The scene disappeared. I'd tripped a breaker.

My father then talked about how the U.S. base in Phu Bai was always under attack when he was there. The base with the helicopters, heavy guns, minefields, razor wire fences, sand bunkers, and artillery. But the VC never attacked the Thai base after that demonstration, and that base didn't have anywhere near the defenses of the U.S. base. They were a much easier target.

I was still stuck on the interrogation. *Two hours!* I couldn't get past how long he had to watch. I sat there, speechless.

"Don't worry," my father said softly. "I'm not going to tell you about the really bad stuff that happened over there."

I've never felt sorrier for anyone.

-atypical-

The pain grew.

The hip surgeon—a leading expert and pioneer in minimally invasive procedures—frowned. He ordered another MRI and another CT scan. He prescribed more medication, more PT. He did not think another surgery would help. The symptomology was atypical, code for he didn't know what was wrong.

A month later, I stood in a forgotten corner of the home goods section at Target, staring at a pile of beige foam seat cushions. It was déjà vu all over again. Back to carrying the cushion everywhere. Back to the awkward interactions and stupid jokes from co-workers, Friday nights on the couch, ice packs and medication, weekends alone, and exhaustion. I bent over, hands on knees, each breath a fight against an avalanche of memories.

To complicate things, the post-surgical groin tightness persisted. After eight months of deep tissue massage, standing for long periods still proved difficult, and sitting remained limited because of the hip. Somehow, I was worse off than before the surgery. I canceled a flight to visit friends in California. I turned down an offer to help a co-worker coach his son's pee-wee basketball team. I shut down my online dating account. Life shrunk.

Cathy did what she could to keep me upright. We scheduled more appointments. I added massage to help with the depression as much as the hip. Between PT and massage, I had several treatments every week. I added acupuncture. I started seeing a chiropractor. Something had to work.

In October of 2010, a sharp stab preceded a sudden weakness below and to the right of my belly button. The area swelled and yelped when I walked or stood, leading to a second rabid dog in my skull jockeying for attention. In the mirror, my reflection squinted back at me as if the world had become a little too bright and loud.

A month later, the hip exploded. No warning. No trigger. Deep, radiating lightning tunneled through the hip and leg. Every position brought agony. I covered the area with anti-inflammatory patches, curled into a ball on the couch, and missed a week of work.

The surgeon ordered another MRI and another cortisone injection into the right hip joint—the same procedure performed before the surgery. The MRI revealed nothing. The shot did nothing. The joint was not the problem. He shared a knowing glance with his office manager, and the supreme confidence I'd once had in him jumped out of the ten-story window.

"I think you might have a sports hernia," he said, as he pressed into my lower abdomen.

"What's a sports hernia?" I asked.

"Torn adductor or abdominal muscle," he said. "We see it every now and then with hip pain. There seems to be a relationship for some patients."

Torn muscle? My head spun.

"Sometimes, if we can fix the groin, the hip pain goes away, too. I'm not an expert, but I know someone who is. He's in Philadelphia."

Philly? Now the room spun.

"Nobody in New York can help me?"

"Not like him," he said.

So, to recap, I started with a sore hip, pain near the bone on the side of the leg. After surgery, the groin muscles felt like they belonged to someone three inches shorter. One of those muscles eventually snapped, and the *new* groin pain had somehow caused the *old* hip pain, which meant time was flowing backwards, the future was the past, and I needed a different surgeon two hours away because nobody in New York City could help. Down the rabbit hole I tumbled.

As I checked out with the receptionist, I gazed up at the wall behind her, covered with pictures of professional football players. Some played for the New York Giants. Others for the Jets or Bills. All were signed with a sharpie ... *Thanks for getting me back on the field, Doc!* Careers requiring hips strong enough to sprint 50 yards and tackle 240-pound monsters, and all I needed was a hip good enough to work an office job. I should've been an easy case. Instead, I was atypical.

-tapes-

While Larry was off in Vietnam, Marie signed up for her first charge card with Barnard's Department Store in downtown Worcester, not really understanding the concept of credit.

"I just kept charging things and sending them to your father. It was wonderful!" she said. "I didn't realize I was going to have to pay it back."

She remembers sending him food, lots of food, and a plastic Christmas tree and bulbs for the holidays. My father said she also sent the traps for the rats in the barracks, but my mother doesn't recall that.

She also used to send him tapes. "Messages on cassette," she said. "Phone calls were rare. We wrote letters too, but it was nice to hear each other's voice. I don't know where we got the idea, maybe from one of the other soldiers."

"What would you say on the tapes?" I asked.

She couldn't remember. Said she'd check the attic to see if they were in some dusty corner, long forgotten. But she did recall one thing my father had recorded. It might have been the last tape he sent her.

"He said that the war wasn't worth one American life."

-procession-

I shuffled back to work and reduced my schedule to 8-hour shifts. December became an endless grind—strategically planned breaks, long lunches laying down in my car. Whenever possible, I snuck out a few minutes early. Every second off my feet was a blessing.

Then, even that was too much. In early 2011, Angela from Human Resources called me into her office.

"You can't keep doing this," she said as I closed the door behind me.

"Doing what?" I glanced at her computer mouse, upside down and on the left side of her keyboard. Down was up. Right was left. How did this woman's mind work?

"Burning up your vacation time," she motioned to the chair in front of her desk. I shook my head.

"I can't lose my job," I said. Someone else had gone in my place on a business trip to Germany. Talks of another promotion had ceased.

"You won't lose your job. You have a medical problem. We have short-term disability, and we're covered by FMLA. Vacation time is for vacations, not lying on a couch."

Almost a year had passed since the pain had boomeranged. Full-time work was no longer an option, even with a makeshift standing station. I limped home at noon each day and iced, chalking up the afternoon to vacation hours on my time sheet. This had been going on for a few weeks.

"Use the benefits we have and get better," she continued. "Get surgery. Get rest. Whatever your doctor thinks you need."

And what if the doctors didn't know?

I'd used up all of my PT visits and covered the groin and hip with anti-inflammatory patches daily, and I was still losing ground each week. As a last resort, the hip surgeon prescribed a Hail Mary and a cortisone injection into the bursar sac behind the hip joint. Maybe it was bursitis. The shot did nothing.

Right before New Year's, I'd driven to Philly for an MRI, followed by an appointment with the groin surgeon. After a 90-minute wait that I wouldn't have tolerated anywhere else I met with the doctor, who kind of reminded me of Mr. Bott.

"Have we met?" he asked, and I nearly jumped. Did Mr. Bott have a brother he never mentioned? No, we determined after a few awkward questions, he did not.

"Okay, let's take a look at what's going on." He snapped on a pair of latex gloves. "Please drop your pants."

He pressed into different points on my groin, probing for painful spots, and I wondered at what moment in his life he knew this was his calling. A few uncomfortable minutes later, he told me I had a sports hernia, a torn abdominal muscle at the pubic bone attachment site. I also had a small inguinal hernia he could fix during the surgery. Ninety-eight percent chance of success for stopping the groin pain, but not sure about the hip.

At checkout, his office manager informed me that the surgery, by the way, wasn't usually covered by insurance. Seven thousand dollars and another $500 dollars for today's visit. I coughed and sputtered.

"I'll think about it," I said.

For the long ride back to New Jersey, I turned on the car radio to drown out the complaining hip (it hated sitting in the car), and I thought about something else.

If I left work, what would bring me back? A groin surgery that somehow healed the hip? More PT? More medication? All seemed unlikely.

Rest might be the answer, I told myself. In the four years since the injury, the only extended break I had from the pain happened after the hip surgery. Maybe the surgery wasn't the reason I got better. Maybe the six weeks of rest did the trick. Maybe.

A month later I left work on short-term disability. Co-workers wished me well, and I told them I'd be back soon, just a few weeks or so, blah blah blah, every word hollow and empty. Nobody knew why I wasn't getting better, and as I exited the parking lot, I worried that I'd never set foot in the building again.

The mid-day drive home should've taken ten minutes, but the New Jersey construction was in full swing at multiple sites. Even in freezing February, the sun low and weak, the sky electric blue, the crews hammered away, dragging the 3-mile commute out over 40 minutes of stop and go. It reminded me of a funeral procession.

IV

-end of tour-

In January of 1969, Richard M. Nixon became the President of the United States. With Vietnam, he promised to achieve "Peace with Honor," hoping to negotiate a settlement that would allow the half million U.S. troops in Vietnam to be withdrawn while still allowing South Vietnam to survive.

Three weeks later, the North Vietnamese and Viet Cong launched a major offensive, attacking U.S. bases all over South Vietnam. Over a thousand Americans were killed. At the same time, South Vietnamese towns and cities were also hit, with the heaviest fighting around Saigon. Eventually, the American artillery and airpower overwhelmed the offensive.

By February, President Nixon was authorizing the bombing of North Vietnamese and Vietcong bases within Cambodia despite government restrictions. At that same time, Larry was on a plane back to the United States, his tour ended.

Marie met him at the airport.

"How was it seeing him after all that time?" I asked my mother.

She considered the question, searching for that 23-year-old again. Her husband had survived the war and they had their whole life ahead of them. A happy memory, I had thought. But her face dropped as the images surfaced, and I immediately regretted the question.

"Oh," she breathed. "He looked … just terrible."

-siege-

Sometimes, I was at Clark University on the rubber floor that made your knees ache for days or Babson with the four-lane track circling above and the high-end fitness center behind the benches. Sometimes, it was my high school gym, small and overflowing with fans and cheerleaders, the bleachers rumbling. I once stepped onto an empty court in the clouds with golden walls and a glass dome ceiling, sunlight bouncing everywhere.

Coaches and players changed too; college teammates, kids from the old neighborhood, relatives long passed. My father enjoyed a regular cameo as an assistant coach or trainer.

The trouble always started during the warm-ups; I couldn't find my uniform; the corridors in the locker room turned and descended, revealing secret passages; or the police would arrive, suspicious of my recent whereabouts. Whatever the reason, I ended up late to the game, lacing up my sneakers at the end of the bench. The other team would score a basket, and the coach would rush over to me, *C'mon! Let's go!* I'd fumble with my shoelaces and have to start over.

When I'd wake, my jaw would ache, and my hands would be balled into fists. The eczema along the right index finger flared, and the knuckles unlocked slowly as if filled with glue. While the shadowy living room materialized, I'd remember, I didn't get into the game. I never got into the game.

"You've had this dream more than once?" the therapist asked, checking her notes.

It was happening all the time.

"What do you think it means?"

"We both know what it means," I said.

"But it helps to verbalize it. To hear yourself say it." She sat back in her chair and sipped from her mug. She didn't care about dreams. She just wanted me to talk.

I remember our first session.

"Are you having any thoughts about causing self-harm or suicide?" she had asked.

I once worked with a heart surgeon, who told me about a rare condition that caused a nerve—I think he said it was in the face, near the cheek—to inexplicably fire pain signals so excruciating and relentless the nerve had to be severed. Otherwise, the patient would commit suicide. He told me this with a touch of sadness as if he'd seen cases, and I felt my insides slip and skitter. A dark, primal secret: given enough pain, we'd all jump ship. Wasn't pain supposed to protect us?

I wasn't anything close to that sort of full-on assault, but the siege was in its fourth year, the pain coursing along the outskirts of my sanity, cutting off my abilities like supply lines: sitting, standing, basketball, dancing, hiking, dating, employment. Sleep was the latest casualty, which meant everything up until then had been a cakewalk.

Three weeks had passed since leaving work, and my recovery plan had failed spectacularly. I rested and stayed horizontal, mixing in walks, errands, and home PT, which somehow led my neck to scream holy bloody murder round the clock. Maybe it was unhappy with all the lying down. I didn't know for sure. In addition, the left hip started hurting again for no apparent reason. Cathy's face went blank when I told her. There isn't enough time in our sessions, she said.

When I slept on my right side, the hip barked after 20 minutes. When I turned over, the left hip flipped out after 30. If I lay on my stomach or back, the neck yelped immediately. I built props and supports with pillows to sleep on the floor, sleep on the couch, or doze in a chair.

Between the scattered naps, I took late night shuffles inside the apartment and peered out into the neighborhood. Dark, lifeless windows stared back at me. *Shh!* they said. The families inside needed their rest. School tomorrow. Day at the office. Maybe a big presentation, then play dates, homework, dinners. and team practices. So much to do!

I avoided the bedroom and its infernal stillness and stayed in the living room with the TV tuned to the weather station. A soothing drone of sun in Phoenix, light rain in Seattle, flurries in Chicago. Something about the movement of weather comforted me. Around 4:00 a.m., I was joined by the faint rumble of traffic to kick off the morning commute. I wondered what poor bastards commuted that early.

Breakfast at seven, which didn't take up enough time, but I couldn't sit for long, anyway. Then rest and maybe a quick nap if I could manage, followed by a trip to the grocery store, a ghost town on weekday mornings. A couple of retirees. Maybe a young mother and her infant in a rush. Her day would disappear before she could blink. By 9:30 I was home again, resting and trying to fill the hours. Daytime TV? Clearly designed to motivate people to get back to work. I read mostly: *The Four Agreements*, Arnaud Desjardins, Wayne Dyer's *Power of Intention,* and other books that said our thoughts created our world. You get what you think about, good or bad. Red Hawk's *Self Observation* challenged me to know thyself weary traveler, and I thought yes, so weary. Lunch at noon, dinner at six, and then dusk slithered in like the villain in a horror movie. I was starving for one good night of sleep, and I'd think about the old Freddy Krueger films, the victims fighting to stay awake. Whatever you do … Don't. Fall. Asleep. No serial killer haunted my dreams, just basketball games and stabbing pain.

But was I suicidal? I had no desire to ingest a handful of pills or jump off a bridge, but other thoughts were creeping around. Like that day the semi ran the red light at some ungodly speed, and I slammed on my brakes, the seat belt digging into my shoulder. I didn't swear or yell. Anger never registered. Instead, a question floated to the surface.

What if I hadn't seen it coming?

And then an answer, gentle and delicate, a thoughtfully wrapped gift.

There would have been peace.

-unclear intentions-

Larry was down to 150 pounds, and I can't remember my father ever weighing less than a buck seventy, even when he was running every day.

With his tour ended, Larry and Marie moved to Miami where he was stationed to finish out the last few months of his service in the Army. There, his stomach issues worsened. Chronic diarrhea. Discolored stools. He was treated for pancreatitis at some point. Was it the malaria medicine? The herbicides? An intestinal infection? My mother couldn't remember the cause. My father said it never went away. He always kept an extra pair of pants and underwear in his car.

Then there were the 52 cysts and their asymptomology. Mysterious passengers with unclear intentions. Doctors examined, poked, and said, "*Hmmm.*" A surgeon said he could biopsy or remove the cysts but was afraid of releasing those intentions into the rest of the body. So, they remained untouched, a reminder of 1968.

Not long after Larry and Marie unpacked their last box in their first home together, the total U.S. combat deaths in Vietnam exceeded the 33,000 soldiers lost in the Korean War.

In November of that year, President Nixon addressed the nation. "Let us understand: North Vietnam cannot humiliate or defeat the United States. Only Americans can do that." (Lindsay, 2015)

-best twenty-

Every now and then I talked with an old friend over the phone.

"Whhasssghoing onn?"

I never said what was going on. I'd sooner chew on a handful of tacks than complain to Steven about a sore hip.

"Whenn'rhe yhou cohmming hhohme?"

I hadn't lived in Millbury in 15 years, but he always referred to it as *home*. Maybe next month, I'd say. Though, there wouldn't be any visits; a 3-hour drive would wreck me for days.

So, we'd settle into comfortable conversation about our Celtics; Paul Pierce, Kevin Garnett, and the other players; who we liked; and who we didn't.

"Trhade Gharnett! He's dhone." One bad game was all it took for Steven to pull the trigger. "Trhade Allen. Hhees terrhible!" Then the Celtics would win five in a row, and the bromance was back on.

Often, the discussions would pull us back to our high school days, and I'd tease him about his antics and Coach Dunham. We'd reminisce about the different gymnasiums and opponents. The tiny gym in West Boylston. The dusty floors in Leicester. Sometimes, we'd talk about summer leagues and sneaking him and Bubba into the WPI gymnasiums.

The conversations lasted maybe 20 minutes. Sometimes, it was the best 20 minutes of my week.

-neural tubes-

In the spring of 1971, Larry and Marie were about to become parents while anecdotes of birth defects and Vietnam Veterans swirled around expecting families like cyclones. Many suspected something to do with exposure to the chemical herbicides, but nobody knew for sure, including doctors and scientists playing catch up to the vast oil slick of dioxins spreading across the globe.

Many years later, those scientists and doctors would prove that yes, there were birth defects caused from exposure to the herbicides, particularly spina bifida (U.S. Department of Veterans Affairs, 2020). The condition resulted in a neural tube defect where the backbone that protects the spinal cord doesn't form and close as it should, often resulting in damage to the spinal cord and nerves. The resulting physical and intellectual disabilities varied greatly, depending upon the location and extent of the damage.

Even without this information, Larry and Marie were very nervous. But in early May I arrived with ten toes and fingers; brown eyes the color of Marie's and the shape of Larry's, a fully formed neural tube and spine, and a winning ticket in my hand.

-commitment-

"What about Father DiOrio?" my mother asked over the phone.

It took a few moments to recall the name. "Is he still around?"

"As far as I know. I think he's in Charlton, or maybe Auburn."

She had called daily since I left work. Bry, my father, and Aunt Jan had checked in regularly, too. Despite my upbeat *everything is fine* reports, they knew that I struggled and was pretty much alone in my struggles.

"He helped me all those years ago," she offered.

A curious story to say the least.

"Just a thought," she said. "You could stay with us."

I considered the drive to Massachusetts. Three hours of sitting. "Yeah, I don't think so. Thank you, anyway."

That would've been the end of it if it weren't for W.H. Murray's famous piece about Commitment, which I'd taped to my refrigerator.

Until one is committed there is hesitancy, the chance to draw back ... (Murray, 1951)

The poem, a handout from a meditation class, grabbed me the first time I read it. Within a week, I'd memorized every line, the words singing hope—*the moment one definitely commits oneself*—as if it all came down to a choice. Stick to a decision, and the Universe met you halfway; God showed up with *a whole stream of events issuing from the decision.*

My eyes focused on the line ... *that the moment one definitely commits oneself, then Providence moves, too.* If Providence was ever going to move for me, my mother's idea might just be it. I'd attended weekly Mass for 30 years. I must've earned some cred with the Catholics.

An online search revealed that Father DiOrio was still around, still preached, and happened to be holding a service that weekend.

March sunlight streamed in through the dining room window of my small, quiet apartment as the sad-eyed priest on the monitor stared back at me. Days were growing longer; spring neared. All I had to do was suck it up and get my sore leg up to see him. Commit.

After another sleepless night, I checked emails the next morning. A single message waited in my inbox from Kristine (Krissy to her friends), the fun, cute athlete I'd dated during graduate school. Just saying hi. She'd found me through Facebook a year ago and lived five minutes from my mother and Paul. In our occasional back-and-forths, we mentioned meeting up during one of my holiday visits to Millbury. It never happened.

But maybe. I fired off a message to her about being in town for the weekend. I could drive up Saturday morning, rest all day at my mother's, and hopefully recover enough for a visit with her that night. The next day, I'd attend Father DiOrio's service.

Krissy was a single mom with two young kids, though, so I had a better chance of winning the lottery than making any last-minute plans together.

-aftermath-

By 1973, Larry and Marie were building the American Dream: a duplex apartment; a jiggly, giggly toddler; a second child on the way; and a small, cuddly poodle mutt puppy named Peppi circling excitedly. Through Marie's connections, Larry landed a job at the Social Security Administration as a claim representative, helping people with their applications for disability and retirement benefits. After a long day at the office, he'd play with Peppi and me and take us to the nearby playground.

That same year, the United States withdrew the last of its forces from Vietnam. Of the nearly three million Americans that had served in war, more than 58,000 had been killed—most of whom were younger than 21-years-old. In addition, 75,000 had been severely disabled.

After I'd gone to bed, my father would pour himself a shot of Southern Comfort, take the butcher's knife from the kitchen drawer, and place it under his pillow. (According to my father, my mother was okay with that if it made him feel safer. According to my mother, she had no idea he was doing this and never would have agreed to it.) Reports occasionally surfaced in the news about a vet waking up in the middle of the night, delirious with flashbacks, accidentally injuring or even killing his family.

"Your mother's a good woman," my father said long after they had divorced. "She really helped me come home."

Two years later, our family moved into the small, brick red ranch two houses up the street from the Bott's. Around that same time, Saigon succumbed to a full-scale invasion of North Vietnamese forces, and civil wars in Laos and Cambodia also ended. All three countries became communist states. One after another they fell, just like a row of dominoes.

-tex-

She wore skinny jeans and a heather green V-neck Celtics t-shirt. While Paul and my mother rushed to greet her, I stood back, trying to connect the dots between past and present versions. The long, permed, dark curls of a college coed had been replaced with a neat, blond bob that ran well above her shoulders. It projected a playful maturity while also accentuating her hair's fine, silken quality.

After hugs with the parents, her hazel eyes found me, and she flashed that same warm-apple-pie smile that hurt too much to think about after we separated. Someone had nicknamed her *Tex* in High School because of that big, easy grin—lips spreading wide, nose scrunching at the bridge, and eyes brightening and crinkling at the corners.

We embraced, and the ache in my heart rivaled that in my hip. Fifteen years ago, we had said our goodbyes on a summer afternoon in a boxy silver Camry. I regretted the decision for some time. She was married and pregnant within 18 months.

Old stories and laughter filled the evening before Paul and my mother went off to bed. I lay on the couch in the small, dim living room. Krissy curled up in the love seat next to the picture window, an afghan covering her legs. The only light came from a shaded table lamp in the far corner. Middi, my parents' loveable corgi mutt, slept on the floor between us, snoring loudly.

"You drove all the way from New Jersey to go to church?" she asked.

"I felt moved by the Holy Spirit," I said.

"Uh, huh," she rolled her eyes. "This has nothing to do with your hip?"

"I came to pray that the surgery goes well. This church specializes in prayers for the groin."

The corners of her mouth crept upwards. "Catholics sure are different."

"Maybe, but can we talk about something else? Your fixation is making me a little nervous."

"I'm not …" Her voice cracked, and she flashed that smile. "I was just asking about church."

The conversation danced around like a ping-pong ball, back and forth in time. The hip. An impending divorce. All these years, yet still easy and comfortable despite the unfortunate topics of conversation.

At some point, we both recognized this, and a bittersweet silence filled the room. Our eyes met, and I glimpsed an accusation —*Your fault, dummy!* My fault we broke up. My fault we didn't marry, buy a home, and have children. She'd been ready.

Instead, I left to follow a path of healing, trying to help the Stevens and Larrys of the world, and she went on to dedicate herself to raising a family. Those decisions led to others which led us back to the living room of my childhood—a single mom, disillusioned about marriage and anxious about the divorce's impact on her children, and a disabled medical engineer struggling to manage each day. Two lives drowning in irony.

Sometime after midnight, she headed home. We promised to keep in touch, and my heart stirred from its slumber like a hibernating bear, catching a hint of spring in its nostrils.

-what would happen-

For ten years, the United States sprayed 20 million gallons of weapons-grade herbicides over 4 ½ million acres of Vietnam lands.

Four and a half million acres is about the size of Connecticut and Rhode Island combined or a little smaller than New Jersey. There are eight million people in New Jersey, so … eight million people and every other living thing and one morning the sky opens up and rains 2,4,5-trichlorophenoxyacetic acid contaminated with varying levels of 2,3,7,8-tetrachlorodibenzo-p-dioxin, designed to create unsustainable hypergrowth in cells. This goes on for ten years intermittently, and the chemicals soak into skin, hair, eyes, hands, breasts, children, grandchildren, dogs, cats, chickens, cattle, lakes, rivers, farms, soil, trees, frogs, and fish.

What would happen? If you answered, "I have absolutely no idea, but it's gotta be fucking terrible!" you'd be correct. That's exactly what happened.

-divine mercy and healing-

The priest entered from behind the altar looking every bit of his 81 years—bald on top with a thick ring of smoke-grey hair around the ears and back of his head. He moved with a slight limp, an apparent hip or knee problem. When he spoke, a welcoming but weary baritone strained to reach the back of the church.

In 1976, the life of Father Ralph Anthony DiOrio Jr. of the Worcester Diocese changed dramatically when he was *openly blessed with the Holistic Charisma of Healings* (Catholic Online, 2006). In non-Catholic speak, people with various illnesses or injuries healed spontaneously in his presence, usually during his services. Sometimes, he prayed. Sometimes he *laid on hands*. Naturally, this caused an uproar within the Catholic leadership, but after much head scratching and a growing list of evidence, the Church reluctantly gave him its blessing to go forth and be an instrument of God's will. Father DiOrio set up camp in Worcester County and went to work.

I'd heard the anecdotes about the *healing priest*, my father telling me one about a mover whose back went out while hauling an air conditioner into the priest's home. The tales fascinated me as the Church taught that Jesus healed cripples, lepers, and the blind; raised Lazarus from the dead; and then told his followers that they could also perform the miraculous if they only believed.

The legend hit closer to home in the late '90s when my mother had a story of her own. At the time, she suffered from acid reflux that turned her throat into hot sand. Sleeping was difficult, and eating had become a precursor to misery. Doctors prescribed medication, but the side effects were worse than the symptoms. After a year of this, she attended one of Father DiOrio's healing services.

"Acid reflux! Someone is being healed of acid reflux right now!" the priest announced as he strolled among the crowd, sprinkling holy water on the attendees. He would sometimes receive a *Word of Knowledge*: someone's name, a symptom, or an article of clothing. Whomever the clue related to, that's who received the healing. *Acid reflux* happened to be broadcasting on that particular day.

That's me, my mother thought, *I have acid reflux!* She felt something in the top of her chest and lower neck—a warm tingling as if filling with sunshine and seltzer. It spread, and the sensation grew stronger, undeniable. She raised her hand.

"I have acid reflux," she said.

"Not anymore!" He pointed at her and grinned. "God has healed you! And you're going to be able to eat pizza again!"

The crowd laughed. She laughed. But he was right. She ate pizza again, among other things that should have set it off, but nothing ever did. Healed on the spot.

If they only believed.

Father Diorio gave a homily about adversity that reminded me of a Coach Dunham halftime pep talk. Midway through, the focus shifted to receiving healings, and I stood up in the back of the church, ready for anything. But then he switched topics, complaining that he'd never experienced a healing himself. He mentioned the pain he was in at that moment and wondered why so many people around him had been healed but not him, apparently as mystified by the whole "gift" as everyone else. He noted that it would be easier to do God's work if God would help keep him upright—*nudge, nudge, hint, hint, God*. Instead, he was like me: doctor appointments, medication, injections. I dropped down onto the chair, mind swimming with doubts.

Eventually, his message found its way back to healings, comebacks, and resiliency, and this rallied him. His shoulders straightened and voice boomed. "And if you get knocked down, pick yourself up off the canvas, and get back in the fight!" He swung his fist across his body.

He then left the podium and returned to the altar. I practically jumped out of my seat. No healing? No laying on of hands?

Next up was the Liturgy of the Eucharist—a sacrament in which bread and wine were transformed into the body and blood of Jesus in remembrance of his sacrifice. The daily miracle was arguably a priest's most important responsibility to his parish and... *who cares! I drove 200 miles! I want my hip to tingle!*

The offering to the church was next, then the peace offering followed by communion—sit-stand-kneel-pray-sit-stand-kneel-pray—and "Our Mass is ended. Go forth in peace and serve one another." The organ kicked in, and everyone sang. Father DiOrio led the altar boys out the back. The song ended.

I waited as the parishioners gathered up their coats and headed for the doors. I waited some more. A woman standing nearby turned to leave when she noticed me.

"It doesn't look like he's going to have people up today," she said. "He must not be feeling it, maybe because he's tired from his service last night."

"Oh." I leaned back against the wall and closed my eyes. The hip was roaring.

"Sometimes, there are delayed healings that happen after a service," she offered.

Sure there were.

-misery-

In 1978, while Steven was making car rides to the Elmwood Street Elementary school miserable for Johnny and me, 40,000 Vietnam veterans filed a class action lawsuit against Dow, Monsanto, and other companies that manufactured the herbicides used in Vietnam (Agent Orange manufacturers are sued on behalf of veterans, 1979).

The plaintiffs cited their illnesses, their wives' miscarriages, and their children's birth defects as grounds for reparations. Everybody was getting sick. The size and complexity of the case was unprecedented and put litigation in limbo for years.

My father was not part of the suit and remained mostly unaware. He wasn't contacted by an attorney until years later; nothing ever came of the call.

-time enough-

So much had changed in the old neighborhood.

John and Lori Healy, and their little Nick, who was growing like a weed, had built a split ranch in place of the (alleged) drug dealer's faded pink singlewide trailer. A town sewer pumping station surrounded by an 8-foot black chain link fence stood where our bike trails once snaked beneath trees. Unfamiliar faces drove by, and Paul leered at the "damn kids" who sped up and down the dead-end road in noisy, loud, rusting cars with missing hubcaps.

A teenager named Crystal often took her pony of a dog, Brody, for afternoon walks. They'd stop and greet my mother and Middi. "He's just a puppy, part St. Bernard," Crystal said one day. Brody's head was the size of a watermelon. The dogs sniffed each other's noses. and Brody's exhale whipped Middi's ears back like a leaf blower. Middi stumbled and barked his displeasure.

I'd been staying at my mother's house, recovering from the pay-out-of-pocket, IRA-wrecking sports hernia surgery. Despite her very sore lower back, my mother had driven me to my condo in New Jersey, then to Philly for the surgery, then back to New Jersey for a night, and then finally to Massachusetts. The woman was an absolute rock.

Per the groin surgeon, I was to walk one mile every day, and no laughing and sneezing for a week, he added with a wink. A three-inch horizontal scar stretched along my lower abdomen and instantaneous crippling pain in my mid-section followed any contraction of core muscles.

So, I shuffled around the old neighborhood, slowly navigating the half-mile cul-de-sac. Sometimes my mother joined me. Sometimes Aunt Jan. The old Bott house by the river, once light blue with white shutters, once a castle filled with adventure, had been painted mustard with walnut trim. Aunt Jan and I didn't care for the coloring, but of course, I harbored a deep bias. Without fail, a memory or two surfaced whenever I passed.

An American flag still flew on the pole next to the garage, but the legendary Bott-on Garden basketball court of champions seemed nothing more than a worn, sloped narrow driveway lined with cracks and a pole sticking out of the ground at the edge nearest the brook. The Botts had been gone for over ten years, trading in the vertically-challenging 3-story home for a custom-built spacious ranch on the other side of town. A friendly, ginger-haired family of five had moved in, and I wondered why they took down the hoop but left the pole.

Despite the frequent walks, I couldn't reconnect with the neighborhood. Too many folks had left. A few parents (now grandparents) would stop in their cars and say hi, followed by an uncomfortable exchange about why I was back home. But none of my friends remained. They'd moved on: married, raising children, juggling careers and family. The years had raced by.

After the walks, I'd stretch out on a small couch on the porch, sometimes reading, sometimes watching the sun work its way around the high-tension wire tower in the backyard, hoping the surgery would work some kind of magic. Grubs had taken out most of the grass near the shed, leaving large unsightly swaths. Smaller brown spots also pocked the lawn at Middi's favorite rest stops. Crabgrass had taken root and started to fill in the dead patches.

-unusual contention-

In the spring of 1983, Dow Chemical Company maintained that at least two years before the United States halted the use of Agent Orange in Vietnam (in 1971), both the Defense Department and the company were aware of evidence indicating that dioxin, a contaminant in the herbicide, might cause birth defects in the children of women exposed to the defoliant. They pointed to a National Cancer Institute study conducted in 1969 showing dioxin caused birth defects in mice (Burnham, 1983).

Dow's statement, in a motion in Federal court to dismiss the lawsuit by Vietnam veterans, appeared to conflict with the company's own assertions that there was no indication that dioxin caused any adverse health effects. The unusual contention by the chemical company that it had early knowledge of the dangers of dioxin was part of a *government contractor defense*; if Dow could prove it manufactured the herbicide according to Defense Department specifications and both parties were aware of the hazards, then the company could not be held liable for any damages (Dow Corporate, 2023).

-impatience-

The hip never improved, but the groin did, which also meant walking improved. Soon after, Cathy solved the neck riddle, which meant sleep was possible. Days separated from nights, resetting circadian time and quieting the insidious, hopeless thoughts.

Three months passed. Did I progress? Maybe, but my *return-to-work goals* had been replaced with *pain management strategies*. Cathy tweaked my exercise program, and we figured out some lifestyle adjustments along with better activity-rest balance to avoid flare-ups. Sitting, standing, and walking felt normal for short periods of time; the key seemed to be rest. With regular extended breaks, I could control the ache. Not very functional, but the nerves were happier.

"Is there anything else we can do for the hip?" I asked the hip surgeon during a follow-up appointment while he moved my leg through a series of action figure poses. "I'm still not able to do much." I didn't push on how the groin surgery was supposed to help the hip. It did nothing.

Hypothesis number three suggested that the problem emanated from the soft tissue surrounding the hip joint. The surgeon sometimes used platelet therapy for soft tissue injuries but wanted another MRI of the hip and groin before making any recommendations.

"Can I try to go back to working part-time?" I asked.

"Let me see what the MRI shows," he said. "If it looks okay, I'll sign the paperwork."

I emailed Beth, asking if I could get a standing station put into my cubicle so I could cut back on sitting time. "Whatever you need," she said.

In a week, I had the MRI, standing station, and HR return-to-work forms. All I needed was the hip surgeon's signature.

He went on vacation.

I waited another week and researched platelet therapy—injections in which the patient's blood was extracted and then spun in a centrifuge, separating the platelets from the plasma. The platelets were then injected into the injury site to help with healing. Tendon and ligament injuries showed the most promise with the experimental procedure. Use the body to heal the body. Something new. Something different. Hope grew and hummed inside of me, yearning to get out.

The hip surgeon returned. I called. No response. I emailed his office. Nothing. Another week passed. Then another. His receptionist said he was busy with a backlog of surgeries and follow-ups and assured me that I was in the queue.

"How long does it take to read a report?" I asked.

My favorite college professor, Professor H., once told me that the path to becoming a successful engineer wasn't that complicated. Yes, you needed to understand the basics—Ohm's Law, Newtons Laws and so on—but after that, you only needed two things: patience and persistence. As he said this, he spread his arms wide, palms up. *Ta da! Can you believe it's that simple?*

In engineering school, he explained, everyone was bright in math and science. We headed out into the world, degrees in hand. We landed in labs, manufacturing plants, and offices, measuring and testing. We took on insane projects—3-mile bridges, artificial hearts, solar powered cars—in which the solutions were hardly apparent.

The basics allowed us to accomplish these wondrous acts. No matter how complex the problem, everything followed the laws of science. Once you knew the laws, it was only a matter of time and energy to determine the solution.

Professor H., maybe in his mid-fifties, had dark, kind eyes, a receding hairline, and a small pot belly. In his younger days, he'd been a competitive distance runner; cross country in college, road races, and marathons. Often, I'd see him running on the school grounds with other professors. I never mistook him for a gazelle, but he was steady and always out there even during the dark and cold New England winters.

Instead of waiting for the hip surgeon, I made an appointment with another doctor in New Jersey who performed platelet injections, a doctor I didn't know. But a co-worker at the office went to him for a knee ligament injury and told me the injection helped. For some reason, that was good enough for me.

-agent orange-

Right around the time the lovely, 13-year-old Kim Anderson was rejecting my offer to "go out," Federal District Judge George C. Pratt Jr. rejected Dow's motion to dismiss the Vietnam Veterans' lawsuit on the grounds of a government contractor defense. He also ordered documents that had been submitted by the companies in connection with that motion to be unsealed.

Those papers proved enlightening. They suggested Dow and other chemical companies had shielded information from the U.S. Government that one of the herbicides, Agent Orange, contained dangerous levels of dioxin (Blumenthal, 1983). In 1965, when the Government was purchasing millions of pounds of Agent Orange, Dow's toxicology director wrote in an internal report that dioxin could be exceptionally toxic to humans. The company's medical director also warned that fatalities had been reported in the literature.

In addition, further concerns were revealed. Dow scientists had some evidence that exposure could lead to serious illness and death. As far back as the 1930's, a pattern of chloracne outbreaks (skin eruptions of blackheads, cysts, and nodules usually on the face) related to dioxin appeared among workers in multiple chemical plants over the years. Dow and other companies shared with one another anxiety over these illnesses and other findings about dioxin but withheld their concerns and some of their scientific information from the Government. The papers also showed that when Dow warned the Defense Department about the dangers of Agent Orange in 1970, military officials declared they were then hearing about the problem for the first time.

-fifth-

I lay on the exam table, stripped down to my underwear, shivering, as cold vinyl pressed against my skin and a healthy flow of frosty air poured into the small room from the vent above. The doctor clipped the x-ray image up against the light box.

"This is where the pain is?" he pointed to the glowing marker on the screen. "Down here?"

I nodded. He reached over and pressed on the spot. I winced. Bingo.

"This is farther down the leg than I thought, but it may be referred pain from your hip," he said. "Before I inject you with platelets, I want to inject a small amount of cortisone and anesthetic into the area to see if I'm in the right site. If the pain is coming from here, it should go away for a little while, and that's where we'll inject the platelets. If not, then we have to find the real source. Understand?"

Four years ago, I was adamant against cortisone shots. I was about to receive my fifth. My last.

"This might hurt a little bit," he said, but I knew the drill, a pinch and then some mild discomfort.

Lightning and acid zipped down the back of my leg, wrapped around the front and drilled through the kneecap.

"Shit!" I shouted, clenching both fists as sweat poured out of me.

He glanced at me before turning back to the syringe, the plunger descending slowly. The storm quieted, replaced with a fullness, a faraway tingling as if the nerves in my leg were sending signals from Fiji.

"Are you okay?" he asked.

I stared at my leg.

"You should be fine," he said. "Come back in two weeks."

The drive home was pleasant, the area numb. Another respite.

The next morning, I woke to the blessed absence of pain and succumbed quickly to hyper-optimism. I sat and ate breakfast, a small miracle. I then sat on the couch and watched TV. Ninety astonishing minutes with only small breaks and zero discomfort. It'd been over a year since that happened.

It's not referred pain! We found the cure! Finally! My thoughts leapt and bounced and tripped over themselves like exuberant children at recess.

An hour later, something at the injection site let go, becoming weak and gelatinous. The leg buckled and a strange tingling spread across the top of my foot.

-unsettling-

The class action lawsuit brought by Vietnam Veterans and their families against the manufacturers of the chemical herbicides was settled hours before it went to trial in May of 1984 for 180 million dollars, a record at the time (Blumenthal, May 8, 1984). Because the plaintiff class was so large, a settlement fund was created and distributed to the veterans and their families through two vehicles: a payment program, which provided cash compensation to disabled veterans and survivors of deceased veterans; and a class assistance program, which provided funds for social services organizations and networks. This plan for distributing the settlement fund was the brainchild of a series of "Fairness Hearings" held in six different locations in the country. Per the terms of the settlement, the chemical companies never admitted culpability (Blumenthal, May 8, 1984).

-middi's room-

A month vanished.

I had zero strength in the hip, glute, and hamstrings, and flexing those muscles felt like being hit with a sledgehammer. In addition, new symptoms joined the old: swelling on the side of the leg and a tingling zip-lining along the shin and foot as if my sciatic nerve had been replaced with an irate electric eel. The cherry on top of this big, steaming pile of potential malpractice was a burgeoning tendon problem. Muscle strains happened daily—a pinch in the wrist lifting a bag of groceries, a sudden jolt along the inside of my knee pushing off a stair, both elbows, the left side of the abdomen, the arch of my left foot, the right shoulder, a big toe. Internet searches suggested several explanations: too many cortisone injections, or maybe a reaction to long-term use of the prescription meds, or maybe a combination, or maybe something else.

The hip surgeon, who finally got back to me, ordered more MRI's, one of the lower back. "Tingling in the leg can be related to a problem with a disc," he said.

The injection hadn't been anywhere near my spine. Regret festered. Why hadn't I waited for him to return? Why did I get the shot? Spinning in circles, no answers to be found.

I had one question. What now?

"You can always stay with us," my mother said over the phone. "Paul and I would love to have you."

She had called as I pulled into the mostly empty parking lot of the Super Foodtown, 15 minutes before the store closing. Since the injection, I shunned big box retailers and supermarkets with their vast floor plans and crowds and snuck into smaller grocers early in the morning or late at night when traffic and checkout lines were at a minimum. I was living in the margins of each day.

"Middi would love to see you too," Mom added before I could respond.

I chuckled. Middi, canine royalty, slept in the guest room, *Middi's Room*. Staying with them would land me with the furry little snore-beast. At least, it wasn't my old bedroom.

A weight lifted with her offer, but Mom and Paul were grandparents in a cozy home covered with pictures of Bry's son, Noah, and great nephews and nieces. Legos, stuffed animals, and board games had been crammed into closets. Squirt guns and soccer balls filled the shed. The house spoke to 6-year-old sleepovers, not middle-aged move-ins.

I recalled the beginning, the doctor with the sausage-thick fingers. A sprain.

I stared at the entrance to the store. My mother was still talking on the phone.

Five years gone. A sprain. A routine diagnosis. Somehow, I had just turned forty. Forty.

I hung up the phone and vomited a mushroom cloud of obscenities at the windshield, the Tai Chi teacher, the surgeon, cortisone, fate, bad luck, the universe. Ragged fury and fire poured out of me, and when that wasn't enough, I yanked on the steering wheel, trying to rip it out of the car. It held fast, so I punched it, and a sharp pain shot through my wrist. *Yousonofabitch!* I hammered away on the dashboard as the cellphone rang.

And then, minutes or maybe only moments later, I was resting my head against the steering wheel, swallowing large breaths. Blinking stabs ding-donged throughout, and my wrist yelped.

The cellphone chimed.

When my breathing finally slowed, I called my mother back and said something about poor reception and dropped calls—and then a very long, uncomfortable … "Yeah, um, anyway, uh, yes, I'll take you up on your offer, um, thanks, Mom, thank you very much."

Sparks cascaded down my shin, and the muscles half-squeezed and protested as I stepped out of the car and shuffled toward the store.

-case study-

My father met Sharon Budzyna in 1986 at a Parents Without Partners Dance.

"She had three guys chasing her at the time," he told me.

My father had a "good soul," so she chose him over the other suitors. Their relationship blossomed with a little help from the Center for Disease Control.

About six months after they started dating, the CDC asked my father to participate in a study of Vietnam Veterans at the Lovelace Clinic in Albuquerque, New Mexico. They offered to pay his room and airfare and give him a $300 stipend. Neither Sharon nor he had ever been to Albuquerque, so he asked her to join him. Sharon, who happened to work in the Infectious Disease Department at St. Vincent's Hospital in Worcester, knew all about the CDC.

"Why does the CDC want to talk to you?" She asked warily. "What's wrong with you?"

"That's what they're going to find out," he half-joked.

"We ended up having a great time," my father told me.

While he was getting interviewed and examined, Sharon sat by the pool and read. In the evenings they went into town and explored. The hot-air balloon festival was happening at that same time. They rode the aerial tramway. They ate at a restaurant high up on one of the mountains that provided a spectacular view of the city. The manager at the restaurant was especially cordial. He made a special chocolate-flavored drink for Sharon.

"When they found out I was a veteran, they treated me like a VIP," my father said. "Sharon was impressed."

That made me wonder about the stories I'd heard about the anti-war protests and treatment of soldiers returning home. "Baby Killers" they had been called, among other things. My father never talked about any of the protests, except once, about Jane Fonda.

"Do you remember anything about the study?" I asked him.

"There was bloodwork and neurological testing. I remember a lot of the vets were experiencing neuropathy," he said. "They were trying to figure out if Agent Orange had caused any lasting health problems."

"How about the interviews?"

"They asked about our experiences in Vietnam and if I was having any health problems. They asked me a lot about my stomach issues from the malaria medicine."

"What about the lumps?"

"Yeah, they looked at the lumps," he said, and then he stopped as another memory took shape. "I remember when we first arrived, the CDC staff told us what to expect for the exams and what not. They only made one request of the Veterans. Please, no crying during the interviews."

-plans-

I used to think it was the therapist's job to heal my hip. Massage it, stretch it, twist it, ice it, whatever it took, and then it would get better. I was wrong.

Cathy was in New Jersey, and I was staying in Massachusetts with a leg that tingled, twitched, and dragged along like a boat anchor.

"You need Susi," the receptionist had declared without hesitation when I called the nearby South County Physical Therapy studio and told her the twisty tale of the hip.

The senior therapist at the studio, Susi Lasewicz greeted me with a handshake, clipboard, and megawatt smile. In the short distance from the waiting area to the exam room, I caught her studying my steps several times.

"How'd I do?" I asked later as we finished up a series of tests.

Her eyes flickered to her notes. "Let's just say I can help."

From day one, a quick and easy connection formed between us, and I became a friend whose story had been flipped, not just her 11:00 patient. The diminutive former gymnast exuded energy and caring, smiling encouragement everywhere and bounding around the studio on legs made of springs. Yet, something more than energy and caring drew us together, something I sensed more than understood.

After a few stubborn weeks, we coaxed some improvements out of the hip in late summer, right around the arrival of high school soccer season.

"You're not going to believe this," she said to me one day as I lay on the treatment table.

"Yeah?"

She leaned over and placed a palm on each of my hip bones, feeling for resistance.

"Morgan got pneumonia in the middle of tryouts," a quiver in her voice.

Susi's teenage daughter had just started her junior year of high school. Tall, intelligent and beautiful, the kid had been dealt a sweet hand except for the curse that followed her every athletic endeavor. I'd never met Morgan, but I'd seen the family portrait on Susi's desk and listened to stories during our sessions.

"She was having a great tryout, too. Breathe, please."

I inhaled.

"She's a skill player, you know. Most of the kids on her team are athletes. Great speed but can't maintain possessions."

"Mmmph," I offered as she tried to push my pelvis through the table.

"She had two one-timers last week with her left foot. Her left foot!"

After spending her sophomore season on the bench, Morgan joined a gym with Susi over the summer and got in the best shape of her life. Then, pneumonia. Not a bad cold or even the flu.

"Roll onto your stomach, please," Susi said.

She checked the muscles along the back of my legs.

"Flex, please," she poked the side of my right glute.

I flexed my glutes.

"Flex," she said again, lightly punching my butt.

"I'm flexing," I said.

"You are? Try harder."

I squeezed harder.

"C'mon, flex!" She poked both glutes, alternating back and forth. "Where are the coconuts?"

"Coconuts?"

"Yeah, you should have two hard coconut-shaped muscles back here. This is all jiggly. Are you really flexing?"

"Yes! I'm clenching my coconuts!"

"Jeesh," she shook her head. "No wonder you can't walk. We've got some work to do."

She began kneading my hamstrings, and her thoughts drifted back to Morgan.

"She's better now, but she has no wind. There's a 2-mile run that she has to finish in under 14 minutes, or she can't make the team." Her voice downshifted. "She gets tired going up and down the stairs."

"I'm sorry."

She forced a smile. "Ah, what can I say? My kids are Plan B kids."

"What's a Plan B kid?"

"You know, Plan B. Nothing comes easy for them. Kyle had the knee thing last year when he was getting more playing time. Morgan catches pneumonia in the middle of tryouts. Plan A doesn't work, so you have to go to Plan B … or Plan C, or whatever. Like me, I'm certainly not on my plan A."

She tapped my right hip. "Are you?"

-scenario-

In 1987, the CDC shut down its $43 million study of health risks posed by Agent Orange, claiming that a lack of military records made it impossible to determine which soldiers had been exposed to the herbicide. By that time, thousands more veterans had come forward to say they had suffered cancers and skin diseases and had fathered children with birth defects as a result of exposure.

The decision to cancel the study drew sharp criticism from Congress and veterans organizations, such as the American Legion, which claimed the study was canceled because of pressure from the White House (Schneider, 1990). A congressional subcommittee was formed to investigate.

The following year, Dr. James Clary, a scientist in the Chemical Weapons Branch of the Air Force who was instrumental in developing some of the technical logistics for the herbicide program (Operation Ranch Hand) in Vietnam (History.com, 2011), wrote to South Dakota Senator Tom Daschle:

> When we (military scientists) initiated the herbicide program in the 1960's, we were aware of the potential for damage due to dioxin contamination in the herbicide. We were even aware that the "military" formulation had a higher dioxin concentration than the "civilian" version due to the lower cost and speed of manufacture. However, because the material was to be used on the "enemy", none of us were overly concerned. We never considered a scenario in which our own personnel would become contaminated with the herbicide. And, if we had, we would have expected our own government to give assistance to veterans so contaminated. (Senate Congressional Record, 1989)

V

-driven-

When I think of Plan Bs, I think of the Botts. Mr. Bott always has something in his hands, and he's in motion: a hammer and nails, heading toward the deck stairs; a brush and paint can, hustling off toward the shed; a socket wrench, crawling underneath the pick-up truck. He's tall, tan, wiry, and bald with long, strong, spidery hands that seem more machine than man.

I remember during his endless back-and-forths to his toolboxes in the garage, he'd sometimes pause to watch us playing our games. After a moment or two, he'd crack a joke or make an amusing observation and then continue on. If he ever caught his boys fighting or arguing, he'd stretch *son* into two syllables, and the world would stop.

Every now and then, it would be just the two of us, and he'd talk about strong minds and strong backs. He'd tell me that I had a strong mind and to keep studying in school. Then, he'd imply that he was more of the strong back type and I'd think he had to have a strong mind to be able to fix so many things. He'd served as a mechanic in the Air Force.

By 1996, Mr. Bott saw what we all saw: Steven had stopped improving. For five years, he'd taken his son to therapy appointments, watching slow steady work turn into gains and a remarkable ability to walk. Then, everything halted with Steven's right leg stronger but very slow and his right hand feeble and even slower. There's so much uncertainty with brain injuries until there isn't. Then, someone says the word *permanent* for the first time, and you can feel the weight of that ten-ton life.

Since Joe's wedding, Steven had wanted a driver's license, as his injury tethered his parents to him. They were trapped alongside him, and that lit a fire under Steven more than anything in his recovery. If he could only drive, he thought, how much would life change for all of them! But his body-brain connection had given everything it had: one strong hand and a right foot too slow to operate the gas and brake pedals. Sometimes, it would take him two hours in the morning to get dressed, go to the bathroom, and eat breakfast.

Mr. Bott, retired by that point, had projects around the house to keep him busy and keep him close to his son. Every morning, before Mrs. Bott went off to work, he'd drive to the St. Brigid Church for the 7:00 Mass. Since the accident, he attended the service daily, often arriving early to say a Rosary, which is something like 50 Hail Mary's, five Our Father's, and five Glory Be's.

The Mass drew only a handful of people, and rumors spread that the poor turnout foreshadowed its demise. Father Markey had been asked if this was a possibility.

"As long as George Bott is here, there will always be a weekday Mass," the priest huffed. "Even if he and I are the only two people in the building."

I sometimes wondered what Mr. Bott prayed for in that empty church. Did he rail against God? Was he furious? Did he pray for healing? I did. I prayed for Steven's healing every day.

But I never asked Mr. Bott about church. That was his business, his time. The rest belonged to Steven.

Five years comprises 1,825 daily Masses. At some point during that fifth year, his son's recovery dead in the water, Mr. Bott picked up the phone book and started calling around about automobile modifications for the disabled.

-obstruction-

In 1990, after a yearlong investigation, a house committee concluded that White House officials during the Reagan administration had controlled and obstructed a federal study of Agent Orange exposure among Vietnam Veterans (Schneider, 1990). The congressional panel said a secret White House strategy to deny federal liability in toxic exposure cases led to the cancellation of a Centers for Disease Control Study in 1987. "The federal government has suppressed or minimized findings of ill health effects among Vietnam veterans that could be linked to Agent Orange exposure," the Agent Orange Coverup report stated.

This bolstered the arguments of two veterans' groups, the American Legion and the Vietnam Veterans of America, who had recently filed a lawsuit seeking to have the CDC resume its study of the health effects of Agent Orange exposure during the Vietnam War.

That same year, my father had been having pain on the right side of his abdomen.

"Fatty liver disease," the specialist said sadly. "Worst I've ever seen."

The liver enzymes and triglycerides were nearly three times the levels of the most advanced case the doctor had come across. My father, who was 44 years old, had five, maybe ten years left to live.

"It's usually a genetic predisposition," the doctor said.

Nobody else in our family had it.

At the time of the diagnosis, I was a freshman in college, Bry a sophomore in high school. My father never said a word to either of us. Looking back, I feel bad I never knew, but in all honesty, I don't think I would have handled it well. That was the same year Steven had his accident.

-free-

If I were to make a list of the worst drivers I've ever encountered, teenage Steven would be number one, with the next three spots occupied by nearsighted family members in their late 80s. I'm fairly certain that most of Steven's family would agree with me.

There was the time he totaled Joe's Mustang while Joe was away on vacation, the time he totaled Mrs. Bott's AMC Wagon, and the car fire. There was also the time Steven, Nelly, and I were heading to Worcester along Millbury St. (the old Route 146) when a fat, fuzzy bumblebee flew in through the driver's side window. Steven freaked, swatting and ducking and swearing at Nelly as if it was his fault. Fifty miles an hour, no seat belts, and Steven swerved between the northbound and southbound lanes, dodging cars as the bee buzzed angrily around us.

In 1996, Mr. Bott took Steven to meet Gary Redding, the owner of Handicap Van Conversions, Inc, a long single-bay garage tucked behind a liquor store on the southbound side of Route 146 in Millbury. The business looked like any other garage, but I can imagine Steven's excitement during that first visit.

"So, you want to drive your car?" Gary had asked.

Driving the Pathfinder? How? It had been six years since his accident, and Steven must have thought he'd slipped into the dream state. After he finished telling Gary all the reasons why he couldn't drive, Gary shrugged and told him how he could.

Steven had two issues as far as Gary was concerned—the right leg and the right hand. The right leg was something Gary saw often, mainly foot amputations in diabetic patients. For some reason the right foot usually suffered the complications. An accommodation existed, and it usually worked; install a pedal to the left of the brake that connected to the gas pedal in the same way driving instructors have a pedal on the passenger's side linked to the brake. The new design would allow Steven to operate both the gas and brake with his left foot. He would need to retrain his thinking, but retraining and rethinking had been his life since the accident.

Gary had also seen his fair share of hand injuries—strokes and amputees—that left the driver with one good hand to control the steering wheel. Once again, a simple yet ingenious adjustment existed. A small knob was bracketed to the steering wheel that Steven could grasp, allowing control with a single hand. The knob was positioned at ten o'clock on the wheel to align with his left hand.

And that was it. With two modifications, Steven could physically operate a vehicle.

Once Gary had made the adjustments to the Pathfinder, Steven was able to reconnect with the occupational therapists at Fairlawn Hospital in Worcester. Several months later, he passed the test for his Massachusetts Class 3 driver's license.

Free.

To run to Goretti's Supermarket in the center of town for a half gallon of milk. To head to the Shaw Middle School and cheer his nephew on in a basketball game. To meet a friend for lunch at the A&D Pizzeria on Elm St. He didn't need to see if his father was busy. He didn't have to ask someone to pick him up. He didn't need his parents to schedule their days around him. They were free. He was free. He got to his car, and he went.

-codify-

In the winter of 1991, President George H. W. Bush signed H.R. 556, The Agent Orange Act, which mandated that diseases associated with Agent Orange and other herbicides including non-Hodgkin's lymphoma, soft tissue sarcomas, and chloracne be treated as the result of wartime service (Young & Reggiani, 1994).

"This legislation relies on science to settle the troubling questions concerning the effect on veterans of exposure to herbicides—such as Agent Orange—used during the Vietnam era," the President declared during the signing (U. S. Government Publishing Office, 1991).

In addition, the act defined a new procedure for determining whether certain diseases were related to Agent Orange exposure. It called upon the National Academy of Sciences to study the scientific evidence concerning the potential health effects of exposure to the herbicides used in Vietnam. It also gave the government the option to proceed with further studies concerning the effects of exposure to herbicides.

A key component of the act was the language surrounding the diseases. *Presumption* and *association* replaced *proof.* If it walked like a duck and quacked like a duck, you didn't need DNA evidence. Instead: log the veterans' health complaints; see if trends were forming outside the norms for the vets; report on the trends; and compensate the vets.

"Why the change in heart?" I asked my father years later. "Hadn't the government just shut down a study?"

"Bush was a veteran. He'd flown in World War II," he said. "Veterans usually look out for each other."

-gulf-

I was never tempted by the Florida sun. In high school, I passed up a Disney trip with my mother and her boyfriend. One of Bry's friends went in my place. I never had a Spring Break during college; winters were for basketball, and I didn't have the money anyway.

For my first trip to the sunshine state, I was 27 years old, and I didn't go to the usual tourist spots, like Miami or Orlando or the Keys. I went to Naples, home of the oldest citizens in the country. Snowbirds and early bird specials. Golf courses, auto dealerships, and churches everywhere where residents "crammed for finals."

I remember John and George Bott sitting to my left on the plane with George's lanky frame spilling into the aisle. I remember catching a flash of lush vibrant green in the small cabin window as we touched down in Tampa. Soon we were standing outside the airport, greeted by warmth, brightness, and a soft breeze. That was not the February we knew.

"Nhottoo bhad, huh?"

Steven limped toward us, wearing sunglasses, a pink floral shirt, and cargo shorts. His skin was the color of coffee.

We complimented his tan.

"And yhouthrhee lhook lhike Cahsper," he said.

We glared. He grinned, a big self-satisfied half-grin seven years in the making.

Pain had driven Steven south. Every December through March his hip, wrist, and neck all recited the refrain of the angry arthritic. His body forecast snowstorms and cold snaps as he upped the anti-inflammatories, counting the hours until Spring. In 1997, he decided he needed a change.

Fortunate in the sense that his accident occurred at work, he had received a settlement that allowed him to live in relative comfort. After enduring several New England winters, relative comfort meant getting the hell out of Millbury after the holidays.

With Gary Redding's gas pedal and wheel knob in place, he drove us from the airport in his gold S80 Volvo—a heavy quiet vehicle with leather interior that felt like riding in a living room. He was renting a 3-bedroom townhouse located in a gated community that boasted wide, even, well-maintained sidewalks and a large pond.

Over the next seven days, an independent life unfolded before me that I didn't think possible. We basked at beaches and went parasailing where we spied dolphins in the crystal aqua below. We dined at Steven's favorite pub and stayed for live blues music. We attended a Red Sox game in Fort Myers, where Steven flirted with a pair of young women sitting in front of us. They were from Massachusetts, prompting him to claim that he'd graduated from WPI. I don't think they believed him, but they appreciated the attention. We also took a tour through the Everglades on a fan boat and visited a gator farm, where we fed alligators marshmallows and learned that the reptiles can use their tails to sort of jump (which caused us to jump).

We also got a glimpse into Steven's routines—his favorite breakfast diner, where the staff and regulars smiled when he entered, his walking routes around the neighborhood, his favorite sports bar. "Steven! What can I get you and your friends?" The bartender greeted us as we sat down.

After seven years, I thought I had a good understanding of his life. I didn't. I was both amazed at how much he could do on his own and sobered by how many challenges he routinely encountered. Things like putting on socks, climbing the ten stairs into his apartment, opening a jar of peanut butter, descending the ten stairs, flossing, or peeling a carrot. Each day I'd pick up on more things I'd taken for granted.

Everything we did moved at Steven's pace, but it was a full week. A full life. And Naples, it seemed, would be good for him. All the restaurants, beaches, and stores he frequented already catered to the elderly and disabled.

The only downside was that Naples experienced the same gravity as the rest of the planet. Everybody trips and stumbles—end tables, stairs, rugs, sidewalks—but most 29-year-olds could balance and adjust somewhat, catching themselves or at least minimizing the damage. Not Steven. If he fell, he dropped like a bag of bricks and he risked the chance of being alone when it happened. No parents. No brothers. No cellphones yet.

On the last night of our vacation, he took us to a restaurant on the beach, claiming he had to show us something on the water afterward. The food was standard pub fare, and we sat at the bar, enjoying our greasy hamburgers. Steven paid the bill.

Unless we made a loser-buys-dinner bet during a card game beforehand, he insisted on buying our meals all week. We were *his* guests, and that meant something. Of course, if he'd won a wager on said card game and was eating for free, he ordered extra appetizers, drinks, and a dessert to go.

"Where to now?" I asked.

"Yhou'll see," he said, limping toward the exit.

George and I exchanged a look—*Mr. Mysterious.*

He led us to a wide pier behind the restaurant that stretched a good hundred yards out into the ocean. The tide was on its way out, and a steady wind carried the moisture from the water below into our faces. Up ahead, a small crowd gathered at the end of the pier.

We passed a couple of fishermen and approached the crowd. A few gulls floated on the waves below. I studied the other bystanders, hoping to follow their gaze in some direction, but nobody seemed to be watching anything. Most were engaged in conversation.

I turned to John. He shrugged.

"Steven, what are we doing?" I asked. "It's getting chilly with this wind. And when the—"

He pointed behind me. I turned and looked off to the west as the bottom of the sun touched the horizon, the colors in the sky starting to shift. He had brought us there to see the sunset.

We New Englanders are hill and valley folk. The sun never sets into water. It rises over the Atlantic in the morning, and disappears each evening behind irregular landscapes filled with pines, office buildings, telephone poles, and highway overpasses. Never something as endless and linear as the Gulf of Mexico.

The crowd stilled as the sun slipped into the water, and both sky and liquid transformed. Shimmering gold spread out across the surface of the darkening gulf. The light blue heavens became streaked with peach and coral. The wind picked up, forcing me to blink as I breathed in the salty air. Minutes passed. My mind grew quiet. Wispy clouds materialized in the western sky, their underbellies glowing fire and scarlet in the dying light. When the sun finally sank beneath the water, indigos and violets from the East crept into the canopy. Time fell away.

Then clapping. The small crowd applauded the sun, water, and sky as if they were actors on a stage. I chuckled. Retirement had occurred a long time ago for most of the audience, but each wore a smile that bore a youthful enthusiasm. Couples hugged and leaned into each other. Friends patted one another on the back.

"Same time tomorrow?" one asked.

"No rain in the forecast until Sunday," another replied.

Steven continued to gaze at the horizon, and I was certain that some part of him had left his body and was out there on the water, maybe running and leaping. This was not the Steven I knew. Something about this community—maybe the wisdom of elders, maybe some connection to the ocean—was sinking in. I'm not sure why, but at that moment, maybe for the first time since his accident, I felt a genuine rush of hope.

He turned to me as we started the walk back to shore. "Prhetty coohl, huh?"

"Yeah," I said. "Pretty cool."

-assassins-

On September 27, 1997, thirteen years after the Vietnam Veteran's case against the herbicide manufacturers was settled, the District Court ordered the Settlement Fund closed, its assets having been fully distributed.

A total of $197 million was paid directly to approximately 52,000 Vietnam Veterans or their surviving family members, an average of $3,800 per veteran.

That same year, the liver specialist was elated to see my father.

"Your numbers have come way down," he said, shocked. They were still high but not lethal. "What are you doing for it?"

My father elaborated. The doctor understood none of it.

"Ayurvedic medicine," my father told me later on.

"What?"

"My doctors couldn't do anything for me. I had to find something," he said. "There were a few doctors out there trying other things, like supplements."

Doctors must've been scrambling to help the veterans and their bizarre symptoms. I remember my father talking a lot about Robert Rowen, a doctor in Alaska who was treating patients with nonconventional methods. No Google or physician websites existed in the 90's. Information was found in support groups, journals, and newsletters.

Thirty years later, and Vietnam was still trying to kill him. I was beginning to understand why he kept the Wolverine action figure on watch at his home. It wasn't the VC he was worried about.

The next assassin would show up a couple of years later.

-spectator-

Most afternoons after physical therapy with Susi were slow and sore. Rest. Ice. Wondering. Hoping. Occasionally, Krissy would ask for help with her kids.

The Raymond E. Shaw Middle School hid atop the tallest hill in Millbury at the end of a half-mile long driveway like a small, secret castle. One fine October day, I stood along the sidelines of the school's soccer field surrounded by a centuries-old forest blooming crimson and sienna and gold. A match played out on the field, and a coach shouted "Cross! Cross! Cross!" from the opposite sideline.

Two girls, arched backs, puffed cheeks, and sparring pony tails raced past me toward the ball in a blur of hunter green and maroon. Chunks of grass popped up from their cleats as they ran.

Kaylee, Krissy's daughter, planted her right foot next to the ball—solid, like a cinder block—while the left half of her body rotated around the pivot in a graceful arc. The defender lunged as Kaylee's leg whipped forward, and the ball disappeared with a heavy *whump!* Both girls toppled to the ground.

Time slowed, and the ball floated up and up and then back down toward a crowd of players jostling for position in front of the goal. We on the sidelines held our collective breath.

Several girls leapt as the ball sailed over their heads and bounced off someone's knee. A Millbury player stepped up to the rebound and launched it off her foot like a rocket past the diving goalie. A *pinggg!* echoed across the grounds as the ball struck the goal post and ricocheted back into play; a defender turned and cleared the ball to the opposite side of the field.

The audience exhaled.

"Nice try, girls!" we shouted

"Well done, Kaylee!" the coach yelled.

Kaylee picked herself up and wiped the dirt off her elbows. Breathing heavily, she walked back up the field, hands on hips. I caught her eye and gave her a thumbs-up. She smiled.

We had met months earlier, standing on opposite sides of her kitchen. I said, "Hi, Kaylee," and she said "hi" back, making brief eye contact before looking down at her feet. I remembered meeting my mother's boyfriends—how unnatural, how unfair. Then she grinned as if someone had just told a joke.

"What is it?" Krissy asked.

Kaylee glanced at me. "He was supposed to have red hair and a goatee. And he was supposed to be ... heavy."

Krissy laughed. "Where did you come up with that?"

"I don't know," she shrugged. "That's just what I thought he looked like."

"Sorry, this is all I got." I rubbed the dark stubble on my head with one hand and slapped my belly with the other. "Just a skinny, bald guy."

She laughed alongside her mom.

Kaylee had made the High School JV team as a seventh grader, partly because she was left-footed, a rarity. I managed cheering her on by alternating between standing and reclining on the ground, propped up on my arms. Reclining moved the pressure from the hips to the lower spine, not a wise tradeoff according to Susi, but what were my options? Handstands?

First-half spectators were mostly grandparents and little brothers and sisters, babysitting time with Grampas and Grammas. Everyone was nice enough, but after several games, questions sat behind their hellos. *Why isn't he at a job? Why is he lying on the ground?*

Half-times were spent in the car, with the seat reclined all the way back. Four years since the injury and every tendon and muscle around the hip still released a deep, *thanks-be-to-God!* sigh that relaxed my whole body every time I got off the leg.

The parents showed up during the second half, dressed for success, and happy to be out of work an hour early. Sean Turner was a regular; he and I had played together during my one year of high school soccer. His daughter, Emma, is very fast I commented to him. We didn't have much else to talk about. Chris Wilbur, another high school teammate, showed up in a shirt and tie. He was the CFO for a nonprofit and was raising four daughters, all of whom were involved in sports. He was also involved in local politics and the School Board, and he ran the youth basketball program with Bubba and Bob Ayotte. He also coached several youth basketball and soccer teams. He did more in one day than I could manage in a week.

After the game, Kaylee walked up to me, her uniform covered in grass stains.

"Nice game," I said.

"Thanks. Where's Mum?"

"Doctor's appointment. She forgot about it this morning and asked me to give you a ride home."

"Oh," a flicker of disappointment. Krissy was the rock in her life. I was still an unknown. "Are you coming over?"

"Yes," I said.

"For dinner?"

"Yes."

Her eyes brightened. Food talk was always safe ground with her.

"What are we having? I'm hun-greee!" As far as I could tell, hungry was her default setting.

"It's a surprise."

"C'mon."

"You'll see. But first, we have to pick up your brother."

She rolled her eyes.

-growing list-

Prostate cancer is the second most common cancer in American men behind only skin cancer. The American Cancer Society (2024) estimates nearly one in eight men will be diagnosed, with 67 being the average age at diagnosis. Most cases are in men 65 and older; it is rare in men under 40.

By 1996, the National Academy of Sciences had found enough evidence of a positive association between prostate cancer and Agent Orange exposure to compel the Veterans Administration to add the illness to the growing list of presumptive diseases related to military service. Years later, further studies would reveal two types of prostate cancer: one slow-moving, the other aggressive. Vietnam Veterans exposed to herbicides were not only more likely to develop prostate cancer but also more likely to develop the deadlier version.

My father was diagnosed in 1999. He was 53.

-mr. ketchup-

Right down the hill from the middle school was the Elmwood St. Elementary School. I pressed the doorbell and waited at the locked front door to be buzzed in, wondering when the uptick in security was added. I hadn't been in the school for ages.

The office was busy with kids and parents coming and going, the same office in which I once spent a whole afternoon for putting gum in a friend's hair. Earlier that morning there would have been a tall man with a pronounced limp and slurred speech, chatting up the admins. He would have been greeted warmly by the kids, bending over to high-five some of them with his left hand.

After signing in, I headed out the back of the building toward the playground. Near the swings, I heard a loud squeal and caught movement out of the corner of my eye. Dropping to one knee to protect the hip, I braced myself as a small, sprinting kindergartener catapulted himself into me

"Josh-eee!" I said as he collided with my chest and hung on like a koala.

"Hi, Chris!" he beamed up at me. "I missed you."

I missed him, too.

If Kaylee was a silvery moon at night, waxing and waning as she tried to make sense of her unsteady world, her younger brother was a bright, warm sun in a cloudless sky. I asked about his day, and he told me about a Woolie Card he got in class for helping another student. He said this as he patted his hair to the side with little chubby fingers, the same hands as his mother.

On our way back to the car, we stopped at the cafeteria to retrieve Josh's backpack. Earlier that day at lunchtime, "Mr. Ketchup" would've been manning the large condiment dispensers at the end of the food service lines, pressing the plunger smoothly with his left hand while the students held their trays in place. The role allowed him to interact with (tease) every kid in the school. Once everyone had been seated, he'd walk from table to table, attending to any conflicts or misbehavior as well as helping to tear open milk cartons. Lunch was his favorite part of the day, just like when he was a student.

The only drawback to working in the cafeteria was the noise as it made it more difficult for the children to understand him. The louder he spoke, the harder it was to enunciate. The students had just begun their journey into the English language, learning in a quiet classroom with the perfect diction of a teacher. When faced with his vocal challenges, children often responded with blank stares.

Just outside the cafeteria was the hallway to the kindergarten classrooms. Sometimes Josh would lead me to his homeroom to show me a class project. Mr. Ketchup never assisted in Josh's class. He helped with the second and third graders, mostly set-up work, handing out papers, paints, glues, and markers while the teacher explained the assignments. If a student struggled, he'd lend a hand, sometimes pulling a chair alongside. If a student had special needs, he'd spend extra time. "Hey, bhuddy! Whahttreyhou mhaking?" He often talked to kids about favorite sports teams or foods and asked about their pets.

Occasionally, he served as the "mhuscle" for students whispering or arguing in the back of the classroom instead of listening to the teacher. Rarely did it require more than walking up behind the kids and clearing his throat.

By 2:30, he was exhausted and would give the teacher a heads-up that he was leaving, always a few minutes ahead of the afternoon rush. He was not good navigating through crowded hallways and parking lots.

"Class, let's say goodbye to Mr. Bott," the teacher would announce.

"Good-bye, Mr. Bott!" they'd shout. Someone always shouted "Goodbye, Mr. Ketchup!"

"Ghoodbye! I'hllsee yhouhtomhorrhow."

He'd limp out of the classroom, the muscles aching in a good way, a welcome weariness after a hard day's work. It was a volunteer position that he treated like a career.

"Are you coming over tonight?" Josh asked as we crossed the parking lot.

"Yes." I tried to have dinner with Krissy and her kids at least once a week.

"And guess what?" I slowed down.

"What?" He stopped, eyes widening.

"Mummy's picking up pizza on the way home for dinner."

"Ooohhhh!" He clapped his hands together.

"Shhhh, it's a secret," I whispered and glanced furtively at my car. "Kaylee's in the front. Whatever you do … don't tell her what we're having for dinner. Okay?"

He nodded vigorously.

As soon as I had him buckled in it was "Kaylee! Guess what's for dinner!"

"Josh, no!" I said.

"Peezaaaaa!" He held both arms above his head and grinned.

Kaylee laughed and peered back at her brother—the little prince sitting triumphantly on his booster seat, his perfect record for blowing secrets still intact.

I once asked Josh if he knew Mr. Ketchup. I can't remember his answer, but Kaylee knew him from her days at the elementary school. Kaylee is eight years older than Josh.

-y2k-

The neurologist pointed with his pen to the image on the monitor.

"See this area around the hypothalamus?" he said. "It's supposed to be smooth."

"It looks rough," my father said.

"Shredded, actually," the neurologist said. "That's the myelin, which can only mean one thing... MS."

In 1999, my father's brain was slowing. He forgot words in mid-sentence, staring off into space, searching. At work he forgot meetings. He forgot to show up to his own presentations. He lost paperwork; he lost folders. He lost someone's disability claim.

My father worked for the Social Security Administration as a Claims Rep for 15 years and then as a Field Rep. He mostly helped people who were sick and injured file applications for disability benefits, which I know was important to him. But if you asked him what he did for work, his response was quite different than his official job description.

"I help people take money from the government," he would say.

After 27 years on the job, he met with his manager, Pete Neuman, and told him about his cognitive struggles. It was time to retire.

Pete, who had served a tour in Vietnam as a helicopter gunner, told my father to use up all his sick and vacation time first and take it easy, try to stretch it out.

By the age of 55 my father was retired with multiple sclerosis, chronic fatigue syndrome, fatty liver disease, a herniated disc, and prostate cancer. He wasn't strong enough for the surgeries or radiation to treat the cancer so he'd been researching unconventional means, trying to keep it from spreading to the nearby lymph nodes and bones.

None of this really mattered, though, because the world was about to end.

News reports were regularly warning us about the Y2K problem— the crashing of global computers. The computers' internal clocks read only the last two digits for the year meaning 1999 would reset to double zeros at midnight on January 1, 2000, possibly crashing systems across multiple industries. Most of us were concerned. My father was more than concerned; something inside him gripped tightly to the disastrous if-then paths that followed.

I think it started with chopped wood. He bought cords of it for the stove in the basement. The grid could go down, he said, we're going to need heat. Then he amassed canned goods and filled up the kitchen cabinets and the pantry. He stockpiled winter clothes, backpacks, knives, and hard liquor for bartering. Dollars might be useless in the new economy.

He also bought ammunition for his pistol and rifle. People might want to steal his SPAM and garbanzo beans. Fear radiated off him like a sunburn, his sleepy blue eyes charged but unfocused, like a full-blown insomniac surviving on caffeine injections.

More and more, conversations with him centered on the coming "difficult times." It was a challenge to talk about family, friends or the weather to pull him back. He was always planning for the next big thing that could go wrong, his mind jumping on any bad news.

When the world didn't end, he made appointments with psychiatrists. They tried OCD medicine which worsened his chronic fatigue and brain fog tenfold. He slept 16 hours a day. He could barely get out of bed.

He went back to the doctors, who were out of ideas. So, he stopped the medicine and chose to live with the compulsions and fear rather than utter exhaustion and confusion. At least, he had energy to get into the ring with everything trying to kill him. He would go down swinging, not sleeping.

"My karma," he would remind me. "I probably ran with Genghis Khan's horde in a previous life."

"Must've been quite the barbarian," I agreed, but I was looking at those lumps on his arms.

In 2002, he married Sharon after 18 years of dating despite claiming on numerous occasions he had no intentions of ever marrying again.

"We already had our kids, raised our families," he once said to me. "We're happy. There's no need to get married."

Nobody attended the ceremony, which was ministered by my mother, a justice of the peace, in her home office. Bry and I didn't even know about it until after the fact.

"Why the change of heart?" I asked my father afterward.

"I wanted to make sure Sharon would get my survivor benefits," he said.

-stubborn-

I peeked over the counter. The leg brace was gone.

"Hey, June, how's the knee?" I asked.

"Much better, thanks. I'm walking now." she said and then picked up the phone and yelled into it, "Chris is here for Susi!"

The door to the treatment area opened, and Heather, a young assistant poked her head out. "Hey Chris, c'mon in," she said.

"Any trips south coming up?" I asked as she led me to a table.

"Next week." She smiled. "He came up two weeks ago so it's my turn to drive down."

June had slipped on the ice last winter, fracturing her kneecap in three places. Heather dated a guy from New Jersey she'd met in college. He'd been job-hunting in Massachusetts so that they could stop the long-distance commute. It's never good when you know intimate life details about the employees at a physical therapy studio. Any therapist will tell you, the longer you're in treatment, the worse your chances for recovery.

I scanned the room full of patients, every face drawing a blank. I knew all the therapists, of course, but there was that sense of being left behind as people came and went, healed and moved on. Susi worked the morning shift, so my appointments were mostly alongside retirees: hip and knee replacements, achy joints predicting weather patterns, and small talk about grandchildren and wintering in Florida. Every now and then I crossed paths with someone my age or younger, but they never stuck around. They got better and moved on.

In the beginning, Susi had performed pure magic with my hip. After a few weeks, I could stand straight and gently flex muscles I thought were dead. Two months later, I could navigate a large grocery store if I leaned on a shopping cart. A few months after that, I added short walks to my daily routine. We high-fived. We laughed. She nicknamed me "Half-Ass" because the hip injury had caused a noticeable weakness in the right glute (coconut). *Don't half-ass it on these exercises!* she often joked. We hit milestone after milestone, certain that a normal life waited right around the corner.

Then something changed. The successes slowed, and weeks became months. I plateaued in late winter, the hip barking no matter what we tried. After that, Susi's customary greeting "Any changes?" lacked its typical spunk and curiosity.

At her recommendation, I met with doc number seven or maybe eight—I'd lost count—a short, energetic man, who practiced in Worcester.

"Well, this is never good," he said, feeling the weight of my 2"-thick medical file.

He then poked and prodded, cracking jokes all the while. When he pulled up the MRI images onto his monitor, he sat in silence, studying them for a solid five minutes.

"Your hip's not going to like me going back in there," he said finally, his eyes still on the screen.

I exhaled, thankful.

He turned to me. "There's really nothing I can do for you that hasn't already been tried."

Another dead end. I nodded and reached for my jeans and shoes. Instead of the usual hand shaking and well wishes, though, he asked a question.

"So, what are you going to do now?"

A flash of anger jumped inside.

Not quit! I thought… like he was obviously doing.

At my job—the one I was about to lose after a year on sick leave—we made devices that brought heart attack patients back from the dead and pulled people through surgeries they never should have survived. A heart surgeon came up with the idea for the device in the '60s, when heart surgeons thought chain-smoking was a good idea.

This was a hip—a ball, a socket, and some muscles. Not a heart or spine. Or Agent Orange Poisoning. Or a traumatic brain injury.

How could I ever quit, watching my father claw and scrape and find ways to stave off terminal diagnoses with little help from doctors? Or watching Steven pick up his ten-ton life and inch forward every day, the days relenting into weeks, the weeks becoming years? Giving up was never an option.

Susi wasn't giving up either. Her latest experiment, Graston, created a small injury which supposedly kickstarted the healing response. It could have just as easily been labeled sadism. I'd tried many crazy things over the years, including a Network Chiropractor who never touched me during the adjustments. Yet, all the pain in my body went away for a week after one appointment, and I thought I was cured. By the third week, the pain had returned along with inexplicable, mind-blowing anxiety and insomnia. Then, there was the kinesiologist, who told me I needed to take manganese. She knew this by having me hold a bottle of manganese supplements in front of my chest with one hand while she pushed against my arm, feeling for resistance. The manganese helped for a week and then caused cramps and tightness throughout my body.

Graston wasn't nearly as mysterious. There were tools—shiny, flat and metallic. Some resembled tongue depressors, others bottle openers. One reminded me of a talon. Common side effects of the treatment included redness, bruising, and swearing.

After the beatings, Susi would help me with stretching and icing, and we'd chat about Morgan, Kyle, Krissy, Kaylee, and Josh. One day, I noticed her limping a little, and she told me about a gymnastics injury from long, long ago that had left her with a sore lower back and leg, a weak glute, and a host of daily struggles, including sitting and sleeping. Some nights she had to sleep on the floor, she said.

I knew right then she would never quit on me.

-bunker-

I once received a Christmas card from The Genesis Club in Worcester, one of my favorite local nonprofits. They provide employment, housing, education, and socialization for people struggling with mental illness. I learned about them from my father, who used to help some of the members with their disability claims.

The image on the card, a joint project by clubhouse members, is a wild impressionist vision of a bird. Streaks of mauve and orange stretch along its slender ivory neck, and a small blue heart sits at the base. The oblong eye is solid black and Saturn, Jupiter, and Neptune orbit inside the head. There's a cherry snood across the beak that melts into gold and cerulean as it descends, and the animal is immersed in a flowing and vivid liquid with words like *Dream, Believe,* and *Imagine* bubbling up from the cosmic soup. Inside, the card is signed by five members, wishing me "joy, love, and a little bit of magic this season...." The artwork is titled *"Nothing is So Broken That It Cannot Be Fixed."* I love it. I wanted to reach out to the artists and let them know. But I didn't. I kept it by my desk and still look at it every now and again.

We used to gather every Christmas Eve at my father's home, the cozy yellow ranch in the center of town, up the road from the wire mill. Bry, Sharon, Aunt Jan, Aunt Barbara, and cousins Rob, Beech and Jamie were regulars. Even Aunt Shelia, who is a Jehovah Witness would join us though she made it clear to us she was there to see her family, not celebrate Christmas. We'd mostly hang out in the living room, which had that special Dad feng shui with the worn and comfy recliner next to the stereo system, the giant blue bean bag, two bookcases filled with fantasy, and science fictions novels on each side of an oversized CRT TV connected to a Sega Genesis with a small box of quirky games he'd found at yard sales. A large picture window, framed in a string of Christmas lights, covered the south wall and looked out onto a tall, sturdy maple in the front yard. I remember one Christmas playing the Dictionary Game, leaving us in stitches. I swear we became closer as a family that evening.

That all stopped, however, at the turn of the century as my father's house transformed, became a gate to another world, maybe Saigon or Phu Bai. First, there were canned goods. Then, backpacks. Then, clothing and more clothing. So much clothing. Winter clothing. Summer clothing. Socks and wool socks. The couch disappeared under a pile of jackets and sleeping bags. The recliner filled with work boots. Boxes of video cassettes were stacked in front of the bookcases and then in front of the TV. The difficult times were coming my father said. We had to be ready.

"Anytime you'd like some help with organizing or cleaning," we all offered, perplexed.

"No, but thank you," he'd say, and there was no gratitude behind his eyes. Instead, a challenge. *Touch my stuff, I dare you.*

"Don't ever try cleaning his home without his permission," a support group counselor once said to me.

"Why not?"

"We once had a woman who acquired boxes and boxes of coffee mugs in her apartment. The Department of Health said she had to get them out because it was a fire hazard. When she couldn't find a place for them, the State had them taken away."

"Seems reasonable," I said.

"She committed suicide."

"Dear God!"

"They identify with what they acquire," the counselor said. "It becomes a part of who they are."

I understood so little.

"It can happen to vets for obvious reasons," she said. "Try to help him be safe. Keep the paths clear to the exits. Keep things away from heaters."

"He bunkers," Bry once said to me. "He was always shot at in the war. Never shot back. He just wants shelter."

Wars don't end.

So, I've been thinking about my Christmas Card and *Nothing So Broken It Cannot Be Fixed*. The word *fixed* digs at me like a pebble in my shoe. There's a destination with that word, a place I'm supposed to be heading. This is probably more inferred than implied, yet I continue to turn it over and over because is that really the case? Really? Are we all going back? Getting fixed? Or cured?

What about broken? What about settling in with that? Of course, try to get better. But what if fixed isn't at the end of the road? Maybe the broken is taking us somewhere else. Or maybe there isn't an end. What if it's okay to be broken? Start there, and see where it goes.

I wonder if the artists would consider shortening the title for me.

Maybe I should just get over it.

-woolie world-

I sometimes brought Josh and my nephew Noah over to the Washington St. Park, the same place I'd played with Steven during the summer league so long ago it felt like someone else's life. At some point, a second court had been built alongside the original, and both were in desperate need of repair. Crab grass and purslane spread out from jagged fissures that stretched across the pavement, and the top of each wooden backboard was missing a sizeable, semi-circular chunk as if a brontosaurus had taken a bite before wandering off. The boys weren't really interested in basketball, though, as they were young and the hoops high. We were there for Woolie World, a magnificent, castle-like playground built in 2008 that drew children from surrounding towns.

"Uncle Chris, watch this!" Noah yelled as he hung from the monkey bar ladder, and "Chris, watch me!" Josh shouted as he slid down the twisty slide. Simple things filled the afternoon as I limped, rested, leaned, and cheered. My attention and driver's license were really all that they required from me, and yet those two things meant so much to the boys. We all need mothers, fathers, grandparents, and other big people taking time and saying things like *Wow!* and *Way to go!* and *How'd you do that?*

After an hour or so, Josh and Noah would tire, and we'd walk back to the parking lot, passing the maroon-and-white *Welcome to Woolie World* sign. lettering. I didn't bother to read the smaller print below until years later.

This 2013 Renovation is a Cooperative Project Between
the Town of Millbury and the Commonwealth of
Massachusetts PARC Program.
Special Thanks to the Many Who Contributed to This
Endeavor Especially:
Allcare Medical Supply, Cameron Inc., Millbury
Fire Department Ladies Auxiliary, Millbury
Firefighters Association,
Millbury Lions Club, Millbury Women's Club, Owen
E. Carrigan Sports Scholarship and Memorial Fund,
R.J. Deveraux Corporation,
St. Charles Hotel, Steven Bott, Wheelabrator Millbury
Inc., Windle Landscaping, Inc.

-kitchen sink-

When my father told the waitress that he'd found the cure for cancer, her smile faltered, and her face blanked as if she'd forgotten one of the specials. While she regained her footing, Sharon gave me a look that said *here we go again*.

For how do you respond to that and the millions of lives and deaths, the hospitals and doctors and scientists, the walkathons and fundraisers? Everywhere leukocytes, lungs, bones, brains, and bladders failing. But in Chili's on Route 12 in Massachusetts, the older gentleman with the walking staff and clothes draping off his thinning frame, ordering the two-for special with two entrees, two sides and a slice of cheesecake (which Sharon ain't sharing!) had apparently figured it out.

Nothing preceding my father's declaration had anything to do with cancer; the waitress had strolled over to take our orders. But in all fairness, was there really any conversation that segued easily into a cure. He took out a pen from one of his many pockets, grabbed a cocktail napkin, and began to write the information for her. The waitress recovered and was ready to work through an ordinary discussion about a miraculous subject. I made a mental note to bump up her tip.

"I have a feeling I'm supposed to tell you about it. You might know someone who's fighting it," my father said mysteriously

Everyone in the restaurant probably knew someone with cancer, I thought and reached for my beer.

The truth was he likely had found something or a combination of somethings that had slowed his prostate cancer. His urologist was amazed—15 years without any radiation, surgery, or chemo treatments. But since he took well over 40 (40!) pills, mostly supplements, each day and adhered to an extremely healthy diet because of food sensitivities, I struggled with his claims of pinpointing any treatment enough to announce to restaurant servers.

"How do you know?" I had asked once. "Maybe it's the Vitamin C. Or the zero-sugar diet. Or the fish oil. Or a combination of several things?"

"If you had cancer, would you take your time?" He'd answered. "Try one medicine, and see if it works? Then another? Or would you throw everything at it at once?"

"The kitchen sink," I agreed. Time was the issue.

Both of us were wary of surgeries or radiation. My grandfather of the kamikaze-dodging, World War II destroyer had endured the cut, burn, and poison treatments of the '90's for his prostate cancer and became impotent, incontinent, and saddled with chronic diarrhea. For 20 years he needed a diaper and catheter and eventually died of a catheter infection. No, thank you very much.

The Chili's waitress took the napkin from my father, looking at the writing as he talked. Sometimes, my father would get hugs from people when he shared his story and information; they'd thank him and tell him about their mother or brother who had just been diagnosed. But our waitress slipped the napkin into her pocket, took our order, and then excused herself, smiling at us with the smile never reaching her eyes.

Sharon and I smiled back. My father sipped at his water, hoping he had helped somebody somewhere.

-old friend-

My last visit to New Jersey took place in February of 2013. All the furniture in my condo had been removed and was sitting in a truck. Movers were hauling stacks of boxes up and down the stairs. With the place mostly empty, the dry, woody scent of the hardwoods filled the space, and the aroma spoke to a new beginning, full of potential.

A young police officer had purchased the unit for an agreed-upon sales price far less than what I paid, but that's what a burst housing bubble does to your finances. I was just grateful to have the mortgage gone after a year on the market, a year since losing my job.

I should've been able to weather the storm, my employer provided long-term disability insurance, but the carrier denied benefits. So, instead of receiving monthly payments to help with bills, I had to hire a lawyer and buckle up for a lengthy appeal process. I suddenly had money problems.

Soon after, I developed sleep problems, lying awake in the endless dark, ruminating about electricity bills, maintenance fees, and property taxes. *What now?* came the question, usually at 2:30 in the morning. The answer was always the same—logical and painful: the injury prevented me from working so I had to cut expenses. I had to cut the biggest expense.

Outside the dining room window, the colossal oak in the backyard towered over the nearby buildings, magnificent even in its winter slumber. Absent were the friends that populated it during the warmer months: the chubby squirrel, bickering jays, and brilliant cardinals. Beyond the tree, a western sky filled with broken clouds struggled to decide between overcast and sunny, and I realized that we'd be long gone before the sun dropped below the horizon. I'd seen my last sunset there. I took a deep breath. Goodbyes are funny things with their random triggers.

The previous night Tara had stopped by. She looked happy and healthy. There was a glow about her—and a diamond on her finger. She'd said yes to a young entrepreneur, and they'd bought a home together. Over the years, she and I had remained friends and kept in touch, but with my move back to Massachusetts, she said that we'd be shifting from good friends to old friends.

As she was getting ready to leave, I grabbed the *Explorer's Guide 50 Hikes in New Jersey: Walks, Hikes, and Backpacking Trips from the Kittatinnies to Cape May* from one of the boxes and handed it to her.

"You should keep it," I said. "I can't use it anymore."

She regarded the book for a moment. Then, her face collapsed, and her eyes fell to the floor. Then, she was wiping away tears, and somehow, I was feeling it, too. Six years had snuck up on us, and that book was so much more than a book. "I'm sorry," I said. "I'm sorry," she said. There was a hug that lingered, and then she was down the stairs, book in hand, the door closing behind her.

In the corner of the dining room sat a box—alone, as if it knew it was different from all the other boxes—and I was certain I'd never seen it before. It was thick-walled and cubic with slots cut into the sides for hands to grab hold off. It had a cover, also very sturdy. It may, in fact, have been the most impressive box I'd ever seen.

Inside, 30 or so green pendaflex hanging folders stood upright as if in a desk drawer. The first folder contained a product specification for a medical device container, and I remembered the factory in Rhode Island where it was made. The VP of Engineering brimmed with pride as we toured the production floor, shouting our conversations over the cacophony of the machines.

I flipped through the rest of the folders: more obsolete drawings and specs in front, blank forms and material guides for plastics toward the back. Beth must have cleaned out my cubicle after the injury, and then they gave me the box without checking it, the job haunting me one last time.

In the last folder, I struck gold, a group photo of the R&D team: Carla, Kevin, and John, who was rumored to sneak naps at his desk; Jim and Jeff, our resident comic book expert, who taught me how to juggle; Dan, our go-to technical swami; and Gary, JoAnne, Maryann, and Beth. In the eight years we worked together, I never found out how many cats

Beth had. Behind the picture, there was a photocopy of Michael Strahan smiling, taken right after Super Bowl XLII, the New York Giants upsetting my beloved Patriots. Kevin and Dan, those bastards, had wallpapered every inch of my cubicle with the picture. God, that was a miserable Monday. I hoped we'd stay in touch.

I pulled out the pictures of Strahan and the R&D group and placed the rest of the folders back into the box. The faint barking within grew louder, reminding me to pace myself or deal with a full-blown hip migraine. I stopped, shifted the weight to my left leg, and waited

"I'm sorry. I'll slow down. I just need a little more time," I don't remember when I started to talking to the hip like it was a loan shark.

According to a new doc, a physiatrist out of the Harvard Medical School System in Boston, I suffered from intractable hip pain, a bulging disk in my lower back, multiple myalgias and myopathies, and radiculopathy in my leg and foot that could be from the disk or could be nerve irritation from the last cortisone injection. To sum up, the pain could be coming from the hip, the back, the soft tissue around the hip, or some combination of all three.

I'd found the doctor the previous year, her optimism bright and infectious during our initial appointments. But after several new medications, a plasma rich platelet injection, and months of active release therapy without any real improvement, her sighs grew heavy and her tone apologetic. More recent appointments covered pain management instead of treatment: medicine, pacing, lifestyle modifications, and physical therapy.

"I'm sorry," she said. "Some people just don't get better." I would learn later that by *some people* she meant the 20 million Americans struggling with chronic pain.

I stayed out of the movers' way until the hip quieted. Then, I rummaged through a garbage bag (ran out of boxes) in the corner of the living room, searching for the yellow folder where I kept personal mementos: old letters; birthday cards and photos; a picture of Krissy and me in our 20s, flashing carefree smiles. I flipped open the folder to drop the Strahan and R&D pictures in front and froze.

There before me sat a faded photocopy of a newspaper clipping from the Millbury-Sutton Chronicle. Riddled with creases, the article had arrived in the mail seven years ago, about a month after its publication in July of 2005.

Attached to the photocopy was a yellow sticky note with a handwritten message.

Thought you'd get a kick out of this. The Legend lives on. - Paul

Below the title, *The Biggest Star on the Field*, was a picture of Steven in mid-stride, throwing a baseball from the pitcher's mound at Fenway Park.

-first pitch-

I've been to Fenway a handful of times, mostly in the outfield seats. Once we sat behind home plate with the players' wives and girlfriends. No clue how we ended up with those tickets. From deep centerfield John, Fred, Mr. Bott and I witnessed a young Texas fireballer named Roger Clemens pitch an absolute gem yet lose 2-1. I don't think Clemens lost another game for the entire season.

The seats at the ballpark are rock hard and small, and some face the wrong direction, Functionally, aesthetically, it's a mess, especially for one of the wealthiest franchises in the league. Yet, it's always jam-packed, standing room only, because we New Englanders love our little sufferings. It's our cramped, antique mothership for everything baseball.

I wasn't there the day Steven stepped up to the mound to throw the ceremonial first pitch. Instead, I was in New Jersey and heard everything secondhand. Paul sent me the article and the photo. Then, I forgot about it.

Looking at Steven's photo in my empty apartment, those last moments in New Jersey, I didn't think about the *How? How did he end up at Fenway?* I didn't try to connect the dots, but if I had, working my way back, I'd start with Chris Wilbur. Chris, father of the four daughters, had a nonprofit career, endless coaching commitments, and zero sleep. He stood along the first base line as Steven limped up to the mound. Earlier that morning, they'd taken a tour of Fenway together, gone behind the Green Monster, and met the scorekeeper, arguably the coolest job in the city.

Months before, Chris, who worked at the Boys & Girls Club, had brought Steven to a charity auction fundraiser. Steven made a generous bid on the *Red Sox Fan Package* and ended up winning a day with the World Series Trophy, a dinner function with Big Papi, a tour of Fenway, and the opportunity to throw out the ceremonial first pitch.

But why had Chris approached Steven about the fundraiser? Not long after Steven began volunteering at the elementary school and became Mr. Ketchup, the holidays rolled around, and some of the kids behaved oddly. Instead of ear-to-ear grins and excited chattering, they shied away from his questions about Santa and presents as their eyes dropped to their desks. This baffled him.

Of her five sons, Mrs. Bott had said that Steven was always the first one up every Christmas morning. Even with his injuries, he still felt young inside during the Holidays.

For the past 70 years, the Worcester Telegram & Gazette has run an annual Santa Fund to buy toys for children in need in Central Massachusetts. One winter, Steven made a donation under the name, SRB-33, his initials followed by Larry Bird's Boston Celtics uniform number. There were lots of threes on the check amount and he might have made the donation on Larry Legend's birthday. We all possessed a somewhat rabid devotion to Larry growing up, and Steven carried that admiration well into adulthood. Anyway, he did it again the next year and the year after, and at some point, someone at the Telegram realized it was the largest individual donation in the fund. That led to an annual interview with Steven, and the articles followed.

Years before he became one of Santa's helpers, Steven had launched his own local charitable endeavor with an assist from Coach Dunham. Shortly after getting his driver's license, Steven approached Coach about an annual scholarship he wanted to fund. Together, they conspired to give an award to the senior basketball player who had improved the most over his high school career. Talent and ability took a back seat to grit, perseverance, and hard work. Each spring, Coach briefed Steven on a list of candidates, Steven made a selection, and Coach announced the winner during the end-of-year awards ceremony. Over the years, Steven increased the scholarship to twice its original amount.

The money, of course, came from the accident settlement, and the terms of the agreement were structured by Mrs. Bott. Up until his fall, Steven was a rash and impulsive barely-adult who had totaled two cars and partied way too hard. Providing him with a lump sum payment up front would have led to financial ruin. Back then, who knew if he would ever be able to work and support himself again? So, when the lawyers convened, Mrs. Bott decided upon monthly distributions that would gradually increase over the course of his life.

Yes, Steven was tempted at times to buy a BMW or a Mercedes but instead bought Volvos for their safety ratings. He then moved on to a Grand Jeep Cherokee, which was higher off the ground and easier for getting in and out. He treated his cranky bones and joints to Florida warmth during the winters and vacationed with his brothers and friends around the country and occasionally to Europe. He liked to eat out with others. He liked to pick up the tab. He used the money to make his life as easy and comfortable as he could. And he used it to help children.

Chris Wilbur, a close friend, who asked Steven to be the godfather to one of his daughters and was connected with everything educational and youth-centered in town, knew this. The fundraiser was an easy ask.

But I didn't think about any of that when I looked at the picture. What I focused on, what grabbed my eye was Steven's left arm.

In the photo, he wears sunglasses, a Red Sox Panama, cargo shorts, sneakers, and a Hawaiian button-up. In the background, I can make out the garage door and Stop & Shop sign in the *triangle*, the deepest part of center field. He's a few steps in front of the pitcher's mound, the rosin bag and game ball lying nearby, and his left arm stretches out before him, angled upwards in mid-throw. Steven is *holding his follow through*, just like the coaches taught us in little league, just like Joe Bott first taught me in pee wee basketball.

Simply put, follow through is the physical manifestation of intention. The brain directs the body and the arm, the arm throws or shoots the ball in a direction, and then the arm holds the position, pointing where it wants the ball to go. It's bizarre how well it works. In the photo, the athletic prodigy of our small-town neighborhood executes it perfectly in front of 30,000 Boston Red Sox fans, his heart probably jackhammering away in his chest.

Steven would tell me afterward that he had practiced throwing in the backyard with Fred for weeks before the game. He was determined to reach home plate, to reach the catcher, to not bounce the ball, he said. Body and brain be damned, he would try like hell to throw a strike.

Which he did.

-presumptions-

We (Dept. of Veterans Affairs) believe that contact with Agent Orange, an herbicide used to clear trees and plants during the Vietnam War, likely causes several illnesses. We refer to these as Agent Orange presumptive diseases. Find out if you can get disability compensation or benefits if you had contact with Agent Orange while serving in the military and now have one or more of the illnesses listed below.

Cancers caused by Agent Orange Exposure

- Bladder cancer
- Chronic B-cell leukemia
- Hodgkin's disease
- Multiple myeloma
- Non-Hodgkin's lymphoma
- Prostate cancer
- Respiratory cancers (including lung cancer)
- Some soft tissue sarcomas

Other illnesses caused by Agent Orange Exposure

- AL amyloidosis
- Chloracne (or other types of acneiform disease like it)

- Diabetes mellitus type 2
- High blood pressure (hypertension)
- Hypothyroidism
- Ischemic heart disease
- Monoclonal gammopathy of undetermined significance (MGUS)
- Parkinsonism
- Parkinson's disease
- Peripheral neuropathy, early onset.
- Porphyria cutanea tarda

You have a presumption of Agent Orange exposure if you meet at least one of these service requirements.

Between January 9, 1962, and May 7, 1975, you must have served for any length of time in at least one of these locations:

- In the Republic of Vietnam, or
- Aboard a U.S. military vessel that operated in the inland waterways of Vietnam, or
- On a vessel operating not more than 12 nautical miles seaward from the demarcation line of the waters of Vietnam and Cambodia

Or you must have served in at least one of these locations that we've added based on the PACT (Promise to Address Comprehensive Toxics) Act:

- Any U.S. or Royal Thai military base in Thailand from January 9, 1962, through June 30, 1976, or
- Laos from December 1, 1965, through September 30, 1969, or
- Cambodia at Mimot or Krek, Kampong Cham Province from April 16, 1969, through April 30, 1969, or
- Guam or American Samoa or in the territorial waters off Guam or American Samoa from January 9, 1962, through July 31, 1980, or
- Johnston Atoll or on a ship that called at Johnston Atoll from January 1, 1972, through September 30, 1977

Or at least one of these must be true for you:

- You served in or near the Korean DMZ for any length of time between September 1, 1967, and August 31, 1971, or

- You served on active duty in a regular Air Force unit location where a C-123 aircraft with traces of Agent Orange was assigned, and had repeated contact with this aircraft due to your flight, ground, or medical duties, or
- You were involved in transporting, testing, storing, or other uses of Agent Orange during your military service, or
- You were assigned as a Reservist to certain flight, ground, or medical crew duties at: Lockbourne/Rickenbacker Air Force Base in Ohio, 1969 to 1986 (906th and 907th Tactical Air Groups or 355th and 356th Tactical Airlift Squadrons); Westover Air Force Base in Massachusetts, 1972 to 1982 (731st Tactical Air Squadron and 74th Aeromedical Evacuation Squadron, or 901st Organizational Maintenance Squadron): or Pittsburgh International Airport in Pennsylvania, 1972 to 1982 (758th Airlift Squadron).

-US Dept. of Veteran Affairs Website, 2023

-endgames-

"Shit," I mutter.

"What?" my father asks.

"That's our exit," I gesture across the two lanes of bumper-to-bumper vehicles to the offramp we're now inching past.

Sharon lets out a worried sigh from the back seat.

New York City boasts ten times the population of Boston with ten times the traffic, but given the choice, I'll drive the grid of Manhattan over the labyrinths of Fenway or Beacon Hill any day of the week.

We'd taken our time, enjoyed a leisurely dinner at a restaurant inside the hospital to wait out the rush hour traffic. It's almost 7:30, and yet here we are, creeping along. Is this mess the tail end of the evening commute? Four lanes merge into two with exits on both sides and road signs everywhere.

"It's okay," I say. "We'll take Storrow Drive and pick up the Pike in Cambridge."

My father nods. Sharon is silent. My hip growls, unhappy with all the damn sitting, but I've muzzled it with a healthy dose of meds. Today was important. I'll deal with the consequences tomorrow.

The hip and I have settled into an uneasy truce. I promise to keep sitting to a minimum, take regular breaks throughout the day, and seek no more aggressive treatments, like needles, and surgeries. In return, I get (mostly) manageable pain, zero flare-ups, and minimal emotional roller coasters. It's an imperfect life with strange workarounds, but there's still plenty to do, like driving my father to the Dana Farber Cancer Institute.

One afternoon, while heading home from PT, my cell rings, and the world grows a little brighter.

"What's up?" I ask.

"Wasss goingh onn?"

"Are you calling to settle your debt? You owe me money, y'know."

"Whaaht? Whaht ary hou tahlking about?"

"Remember the March Madness pool?" Steven was eliminated early when his projected winner lost in the second round.

Silence.

"Five bucks," I say.

"Whadddyhou mhean?"

"Do you want me to come get it now? I've got some time."

"How duyuh know I din-nt whin? Ihsn't the tounnhanamn still goingohn?"

All these years, and it still takes an extra second to figure out his words, every syllable an effort. Strangers sometimes confuse his dysarthria with intoxication.

"You were supposed to pay up front to enter the pool, but I cut you a break."

"Ifftitss still ghoing ohn thenhown I lhose?"

"I also took a photo of your picks so there wouldn't be any confusion... like last year."

He snorts. Last year he claimed I misread his picks. That was after he offered me ten dollars to switch his teams halfway through the tournament.

"Whoss kheepinscohre?"

"The website. It always does."

"Ahnd it's nheverwrhongg?"

"You can check the scores if you think there's an error."

"Yhou dhon't know, that's whahtIthought!"

"It's never b-"

"Yhoudhon't know! Yhoudhon't ..." He erupts into hysterics.

I approach a traffic light, just switching to red.

"Cohmeghet yhour dhamnmhoney. Hih'll bhehhere," he says and hangs up the phone.

There are hospitals, and then there is Dana Farber, possibly mankind's best defense against all things cancer. Reception was a checkpoint. A masked admin eyed us suspiciously for sniffles and coughs. Everywhere we turned, signs instructed us to wash our hands and wear a mask for God's sakes if we presented any cold-like symptoms (this was pre-Covid, mind you). The waiting room shone spotless and carried a whiff of industrial cleaner.

The 7th floor windows faced south and west. No skyline across Mattapan and Jamaica Plains, and the sun blazed bright in a cloudless sky. Nothing out there to distract me from my thoughts, from the reason we were there.

"It spread into the bones and lungs," my father had said, disappointed. "I screwed up, used the wrong supplement," which he believed counteracted another supplement he'd been taking since the initial diagnosis.

You didn't screw up anything! This isn't your fault!

Back at his house, a list of prostate-specific antigen test scores over a period of 12 years hangs on a kitchen cabinet next to the refrigerator like a family recipe for grandma's lasagna. The lowest number is seven, the highest 55. The 55 is underlined.

"Bone cancer is a tough way to go," he said. "My karma."

Steven lives in a bright and spacious single story 3-bed, 3-bath modern ranch in one of the newer developments in Millbury. The rooms are tastefully furnished—Pier 1 maybe—and a stone veneer covers the front of the house. Except for an unfinished empty basement and a mostly unused room above the 2-car garage, there are minimal stairs to navigate. A small, well-manicured yard surrounds the front and sides of the house and 6' white vinyl fencing encloses an even smaller backyard covered entirely in artificial turf.

I knock on the front door.

"Cohmme ihnn!"

Steven purchased the house at a modest price after the market crashed in 2009. In addition to the easy maintenance and few stairs, his home also sits a convenient half mile from his parents. Mr. and Mrs. Bott make daily trips back and forth for errands, meals, and chores, helping their son and continuing to be a vital part of his life. At night, though, Steven sleeps in *his* bed under *his* roof, paying *his* mortgage.

I enter the front hallway, closing the door softly behind me. Little feet thump quickly across a floor from somewhere inside. I follow the sound into the living room. There are no signs of life except for a small leg in lime green sweatpants sticking out from under a pile of pillows on the couch. Cecilia!

Cecilia Ann, or CC to her family, had inherited the strawberry blonde hair and blue eyes of her mother and the unyielding stubbornness of her father, Steven.

CC's mom grew up in Millbury and had known Steven before the accident. They connected online and began dating shortly after. Cohabitation, engagement, and pregnancy all followed, but things didn't work out before marriage, and they agreed upon a shared custody arrangement for CC.

"I wonder where CC is," I announce to the living room.

Silence.

"Is she in here?"

The first time I met CC, we talked and played, and she told me about a butterfly that landed on her hand and how "Bapoo" (Mr. Bott) liked to tickle her belly. Typical adorable kid stuff. I didn't think much of it as I usually got along well with children, but Steven said it'd been an act of

God that she'd spoken to me. From that day forward, she hid from me like she hid from everyone else.

This was a problem. Steven needed help with childcare, and CC wasn't comfortable with anyone besides Mr. and Mrs. Bott.

I decide to let CC be. She'll get used to me someday.

"Hello?" I yell to the house.

"Cohmmng!"

Steven limps into the living room, wearing white Nike cross trainers, black nylon athletic pants, and a bright red long-sleeved t-shirt with Elmo's face covering his entire torso.

I stare at his shirt. "Am I supposed to tickle you?"

"Ffu-" He stops, noticing the little leg. "Thahnk yu! It's CC's favrite. Rhight, CC? Whohs yhur favrite Ses-meee Street character? Issit Elmho?"

The leg twitches.

"Everyone loves Elmo," I say. "Some more than others."

He watches the couch for a few more moments. "Nhot todhay, Ighuess," he sighs.

"She'll grow out of it," I offer, "And before you know it she'll be asking you for the car keys to meet some boys up at the mall."

"Oh, Ghod!" he groans. "Ahnd then uh cahn shhoot me!"

The oncologist was friendly. She had met my father 20 years earlier when my father was first diagnosed. She was noticeably surprised and pleased that he was still living without any conventional treatments. She asked about his supplements, and I got the sense she liked my father and was rooting for him, but she reigned in her enthusiasm after a few minutes.

The medicine she recommended, a very low dose chemotherapy, would starve the cancer of testosterone. My father would need to take prednisone to offset something from the chemo, but I forgot what. They could see him for regular checkups at a clinic in Milford much closer to his home. Everything was on the table. Nothing held back. Words were chosen carefully. It wasn't so much about dying; death was certain. What was unclear was how much longer they could extend his life.

"Two years, typically" she said. "Sometimes, patients can get more."

Steven leads me outside, stopping on his front step so he can peek back in to check on his daughter. She is his world, as well as his greatest source of anxiety. *CC isn't sleeping well. CC's cranky. CC's giving me a hard time.* If it weren't for Mr. and Mrs. Bott's help, I don't know how he'd manage.

A few seconds after the screen door closes, little feet scamper across the living room.

"So, about the five bucks," I hold out my hand. "I need to give it to the winner."

"Whinnher? Who whon?"

"George."

"Gheorge?" He reaches into his pocket and pulls out a roll of bills.

"Yeah, he had Kentucky going further than anyone else." George has been a Kentucky fan for as long as I can remember.

"Gheorge, my brhother?"

"Yes, George, your brother."

"Gheorge?"

I sigh.

He smirks and slowly flattens his roll of bills, peeling off singles.

"Dhid I bheat Phaul?"

My mother and Paul live five minutes away. When he doesn't have CC, Steven stops by to visit. He remains close to both, and every so often spends a day at one of the Connecticut casinos with Paul.

"Yes, you beat Paul."

"I dhid? Yhah! Thaht's ahll thaht rheally mhatters!"

He hands me the money. Three dollars.

"It's five dollars for the entry fee, not three."

He glances at the front door and then shuffles a few feet away from it, pulling me with him.

"Hiff Gheorghe whunn then this isss ahll yhou are ghetting," he speaks quickly, and his left hand gesticulates wildly. "nnnatgam shaaFhkingcheahtrd fh hehntttghngandfafngg…"

I consider myself somewhat of an expert on Steven's unique dialect, and I have no idea what he's saying.

"Can I quote you on that?"

"Fhcking chheehtr," he slides the rest of the money back into his pocket. "Thahts ahll hee's ghettng!"

I take the three dollars. "You know you're a—"

The front door opens, and CC appears. She stops in the doorway and frowns at me.

"Hi, Honey!" Steven says sweetly as I pretend to be fascinated by the dollar bills.

"Daddy, I wanna juicy," CC says.

She takes a cautious step, her attention still on me.

"A jhuicey?" he replies. "Okay. Ih'm cohmingg."

He turns to me. "Hive ghotto gho."

"Yes, you do," I say. "I'll be sure to tell George what you said when I give him his winnings."

He flashes his half-smile, eyes twinkling.

Two years means two summers, two birthdays, and two Thanksgivings. That feels heavy and fast, like a locomotive. I should be paying more attention, I keep thinking. It's going to fly by, and he'll be gone before I can blink.

Storrow Drive follows the Charles River along the Boston side, across the water from Cambridge. It's a lively, windy road with impressive views as the river provides a break from the claustrophobia of the skyscrapers. The traffic zips along, so I'm locked in and white knuckled, hands at ten and two, surrounded by "Massholes," some of the intentionally worst drivers in the country. Then, the cars stop on a dime for no apparent reason as only Boston traffic can do.

To our right, the last light of the day illuminates the city skyline in a magical shadowy palette. Every shade of purple and blue glows dark amid the golden remnants of a sun that's abandoned the western sky. Small beacons of light—windows and streetlamps—come alive across the cityscape to fend off the coming night.

"Wow," Sharon says softly from the back seat.

My father is quiet as he considers the skyline and, for the first time in my life, I'm grateful for Boston's infernal traffic as this will likely be his last trip to the city.

When the cars start to move again, I think about the cribbage games my father and I will play these final two years. Winner gets a buck, two dollars for a skunk. We'll talk about Kaylee, Noah, and Josh. We'll discuss generators, sleeping bags, and workarounds I'm learning for the hip. When he wins, he can go home and brag to Sharon, maybe buy her something chocolate that will earn him a smile and a kiss on the cheek. The old poker player still has something in the tank, he'll tell her. He'll be 75 in two years. 75. An ordinary lifespan if you weren't privy to the details.

CC runs up to Steven and grabs his left hand, securing two of his fingers within her tiny palm. Last year he developed trigger finger in the hand, requiring a surgery and a cast for six weeks.

CC takes three-four-five little steps to every one of Steven's sweeping, unsteady limps. Two years ago, he stepped wrong and snapped his right ankle and tibia, requiring a line of percutaneous pins to set the bones back together. Mrs. Bott had to move in with him for several months.

Medical procedures are popping up like perennials, and I can only imagine the endgame of the early arthritis in his hand, neck, hip and back, his unnatural biomechanics likely accelerating the wear throughout his middle-aged body.

But I'm not going to think about his life ten years into the future because at that moment, right then, I've never seen him happier.

"It is something," my father says, his gaze off toward the darkening horizon. "All of this."

The traffic moves smoothly now as the exit approaches. I exhale, ease my grip on the steering wheel, and let what's left of the day back in.

-just to be clear-

During the (Vietnam) war, Dow, Monsanto and other companies were compelled by the U.S. government to produce Agent Orange under the U.S. Defense Production Act of 1950. The government strictly controlled the transport, storage, use, and the specifications to which Agent Orange was to be manufactured exclusively for the military.

The U.S. courts have consistently ruled that Dow and the other manufacturers bear no responsibility for the development and use of Agent Orange during the Vietnam War, and have dismissed all legal claims to the contrary. Moreover, decades of study relating to Agent Orange have not established a causal link to any diseases, birth defects or other transgenerational effects. Notably, the extensive epidemiological study of veterans who were most exposed to Agent Orange does not show that such exposure causes cancer or other serious illnesses.

-Dow Website, 2023

Larry Richards died on April 9, 2025.

Reference List

Agent Orange manufacturers are sued on behalf of veterans. (November 28, 1979). *The New York Times*, Section A, p. 16.

American Cancer Society. (January 19, 2024). Key statistics for prostate cancer. Retrieved from https://www.cancer.org/cancer/types/prostate-cancer/about/key-statistics.html

American Rhetoric. (October 3, 2010). *Martin Luther King Jr.,* Beyond Vietnam—A Time to Break Silence, *delivered 4 April 1967, Riverside Church, New York City.* Retrieved May 15, 2020 from https://www.americanrhetoric.com/speeches/mlkatimetobreaksilence.htm.

Blumenthal, R. (July 6, 1983). Files show dioxin makers knew of hazards. *The New York Times*, Section A, p. 1.

Blumenthal, R. (May 8, 1984). Veterans accept $180 million pact on Agent Orange. *The New York Times.*

Buckingham, W. A. (1982). *The Air Force and herbicides in Southeast Asia 1961-1971.* Office of Air Force History, United States Air Force, Washington, DC. Retrieved May 18, 2020 from https://www.scirp.org/journal/paperinformation?paperid=11560

Burnham, D. (May 5, 1983). Dow says U.S. knew dioxin peril of Agent Orange. *The New York Times*, Section A, p. 18.

Catholic Online. (August 30, 2006). *Noted preacher and healer, Rev. Ralph DiOrio, brings world renowned healing ministry to the Cox Cent.* Retrieved March 10, 2025 from https://www.catholic.org/prwire/headline.php?ID=2126.

Dow Corporate. (May 21, 2023). *Agent Orange.* Retrieved from https://corporate.dow.com/en-us/about/legal/issues/agent-orange.html

Dux, J., & Young, J. (1904). *Veterans and Agent Orange: Health effects of herbicides used in Vietnam.* National Academies Press.

History.com. (October 29, 2009). TET Offensive. *A&E Television Networks.* Retrieved May 17, 2020 from https://www.history.com/topics/vietnam-war/vietnam-war-timelin

History.com. (November 16, 2009). *Muhammad Ali refuses army induction.* A&E Television Networks. Retrieved May 16, 2020 from https://www.history.com/this-day-in-history/muhammad-ali-refuses-army-induction.

History.com. (August 2, 2011). Operation Ranch Hand. *A&E Television Networks.* Retrieved July 24, 2020 from https://www.history.com/topics/vietnam-war/agent-orange-1

Lindsay, J. M. (April 30, 2015). *The water's edge: The Vietnam War in forty quotes.* Council on Foreign Relations. Retrieved July 24, 2020, from https://www.cfr.org/blog/vietnam-war-forty-quotes.

Makers of defoliants sued on behalf of Vietnam veterans. (February 3, 1979). *The New York Times,* p. 17.

Merry, R. W. (2012). Conkrite's Vietnam blunder. *The National Interest.* Downloaded from https://nationalinterest.org/feature/cronkites-vietnam-blunder-7185.

National Archives. (July 24, 2020). *Vietnam War U.S. military fatal casualty statistics.* Retrieved March 10, 2025 from https://www.archives.gov/research/military/vietnam-war/casualty-statistics.

Richard Nixon Foundation. (September 2, 2017). *President Richard Nixon's 14 addresses to the Nation about Vietnam.* Retrieved March 10, 2025 from https://www.nixonfoundation.org/2017/09/president-richard-nixons-14-addresses-nation-vietnam/.

Schneider, K. (August 10, 1990). Agent Orange study was obstructed, panel says. *The New York Times.*

Senate Congressional Record. (November 21, 1989). Agent Orange: Ten years of struggle. *National Agricultural Library.* Retrieved March 10, 2025 from https://www.nal.usda.gov/exhibits/speccoll/files/original/b9290cd678909240fd2b3ef1505d2d96.pdf.

U.S. Department of Veterans Affairs. (July 27, 2020). Spina bifida and Agent Orange. Retrieved from https://www.publichealth.va.gov/exposures/agentorange/birth-defects/spina-bifida.asp.

U.S. Department of Veterans Affairs. (May 20, 2023). Agent Orange exposure and disability compensation. Retrieved March 10, 2025 from https://www.va.gov/disability/eligibility/hazardous-materials-exposure/agent-orange.

U.S. Department of Veterans Affairs. (January 17, 2024). Agent Orange Settlement Fund. Retrieved March 10, 2025 from https://www.benefits.va.gov/COMPENSATION/claims-postservice-agent_orange-settlement-settlementFund.asp.

U.S. Government Publishing Office. (1991). Public papers of the presidents of the United States: George H. W. Bush (Book I, pp. 114-115). Retrieved from https://www.govinfo.gov/content/pkg/PPP-1991-book1/html/PPP-1991-book1-doc-pg114.htm>.

U.S. House of Representatives, Committee on the Judiciary. (1974). *Impeachment of Richard M. Nixon, President of the United States: The final report of the Committee on the Judiciary, House of Representatives, pursuant to H. Res. 803, a resolution authorizing and directing the Committee on the Judiciary to investigate whether sufficient grounds exist for the House of Representatives to exercise its constitutional power to impeach Richard M. Nixon, President of the United States.* U.S. Government Printing Office.

Young, J., & Reggiani, E. (1994). *Veterans and Agent Orange: Health effects of herbicides used in Vietnam.* National Academies Press.

www.ingramcontent.com/pod-product-compliance
Lightning Source LLC
Chambersburg PA
CBHW070620100426
42744CB00006B/552